TEACH YOURSELF BOOKS

ELECTRICITY IN THE HOUSE

Today, no home is complete without an adequate electricity supply designed to serve all purposes, and this book aims to provide an insight into the electrical installation and equipment of a modern house. It begins by explaining the effects of electricity and how it is supplied to the home. Methods of distribution are next discussed, including circuits, wiring, fittings and switches. Following chapters cover electric lighting and heating, the layout of the installation, electric motors and motorised appliances, batteries and generators, radio and telecommunications, and, finally, electrical safety. Fully illustrated and incorporating metric (SI) units throughout, this book will be valuable not only to the householder but also to the student of electrical installation.

TEACH YOURSELF BOOKS

ELECTRICITY IN
THE HOUSE

J. E. Macfarlane
B.Sc.(Eng.), M.I.E.E., A.M.I.Mech.E.

Revised and enlarged by
G. Davidson
C.Eng., F.I.E.E., M.Cons.E.

ST. PAUL'S HOUSE   WARWICK LANE   LONDON EC4P 4AH

First printed 1945
Second edition 1958
Third edition 1973
Second inpression 1974

ISBN 0 340 15249 4

Printed and bound in Great Britain for
The English Universities Press Ltd
by Hazell Watson & Viney Ltd, Aylesbury, Bucks

Contents

voltage drop calculations—effect of heavy load on sub-circuit—heat and power circuits—the ring circuit—lighting circuits—diversity—installation testing—earthing for safety—earth-loop impedance —earth-leakage circuit-breaker

Preface to 1973 Edition

So much change and development in the utilisation of electricity has taken place since the first edition of this work was published that it is most necessary to revise it but extremely difficult to avoid unduly increasing the volume and complexity of its contents without violating the Author's original intentions. A further problem has been to keep pace with and, in places, forestall changes in the transition from Imperial standards to international standards of measurement that is taking place and is still far from being complete. The need to introduce metrication consistently throughout the revision has been treated equally in importance with the modernisation of techniques, practices and equipment, but many manufacturers have not yet brought their products into line with this principle, so rather than have a medley of units throughout the work a bold attempt has been made to convert Imperial to Metric sizes even where a manufacturer has not issued the metric version of his equipment.

This work is intended for students of all ages who wish to gain some insight into the electrical installation and equipment in a modern house, without having to refer to larger works of a more technical nature on electrical engineering practice, and it is hoped that the contents will prove useful and interesting to students of Building and Architecture and junior classes in electrical installation work, besides all those otherwise interested in domestic electrical services.

Thanks are also expressed to numerous friends and organisations for their help and suggestions in preparing the original work, including the Institution of Electrical Engineers and, in the revision, the Association of Supervising and Executive Engineers, together with all the manufacturers, individually acknowledged in the text, who have kindly provided material for use in the illustrations.

Reference is made to various British Standards in the text and these can be obtained from the British Standards Institution, 101–113, Pentonville Road, London N.1.

G.D.
London

1 Electricity, Its Effects and Measurement

In this country we are fortunate in having a good supply of pure water, and it is one of the public utilities we take for granted. We have now reached a similar stage with electricity supply and, with the rapid development of atomic power and electricity becoming more competitive with other fuels, no house will be complete without an adequate electrical installation designed to serve all purposes.

Electricity is a public utility that must be considered in conjunction with plans for any building, whether it is the humblest cottage or a block of luxury flats. Light, airy houses will always be popular, and novel methods of construction are being devised to incorporate all amenities.

Electricity is an essential service in the modern house and due consideration must be given to its inclusion to give a harmonious whole, not as an item to be added as an after-thought. Much progress has been made since the last war, but developments in the years to come will be greater still.

As a labour-saving agent electricity is clean, convenient and easily controllable. It is essential that electrical installations should be fitted by competent electricians, and such important work should not be given to any odd handyman who can use a pair of pliers and a bit of flexible wire. The National Inspection Council for Electrical Installation Contracting has been in operation for a number of years and inclusion on its Roll of Approved Contractors is a guarantee of good workmanship; membership of the Electrical Contractors Association is also an indication of good repute.

The supply of electrical energy is the responsibility of a

national undertaking with fourteen Area Boards and an
advisory body, the Electricity Council. The Government
Department concerned is the Department of Trade and
Industry, which administers the Electricity (Supply) Acts
1882 to 1936, other Electricity Acts up to 1957 and Electricity
Regulations up to 1968.

Effects of an electric current

We are not concerned with what electricity actually is—
though the electron theory is established on our present
experience—but how it can be used in the service of man. Our
concern is with the effects and how electricity can be employed
in a useful and efficient manner. The inventive genius of man
has devised many machines to enable him to carry out tasks
beyond the capabilities of his own pair of hands, from the
primitive wedge and roller used to build the Pyramids to the
mighty turbines in a modern power station. The effects of an
electric current are worth knowing and are given in the
following sections.

Heating effect

When an electric current flows in a conductor, heat is
produced. The electric fire is a common example, in which
the element, made of a special sort of wire, is raised to a red
heat. The electric lamp depends on this effect, as the filament
is so hot that it is incandescent and emits light. These are
useful applications, but if a wire carries too great a current
for its size, then dangerous overheating will result, with a
consequent fire risk, which must be guarded against.

Magnetic effect

Everyone is familiar with the horseshoe magnet, which will
pick up iron nails, pen nibs and small pieces of steel. The
magnetic compass (see Fig. 1) consists of a magnetised
needle which is mounted on a pivot and points in the direction

of the magnetic meridian. When a wire is carrying an electric current, it is surrounded by a magnetic field and behaves in a similar way to a magnet. If the wire is wound up into a coil, a very much stronger effect is produced, which can be made greater still by winding the wire on an iron core. But this large magnetic effect is only present while the current is flowing, so this property is employed to obtain a magnetic field when and as required, and the combination is called an 'electromagnet'. Special hard steel can be given permanent magnetism by such a method, i.e. it remains magnetised when the electric current in the coil is discontinued. On the other hand,

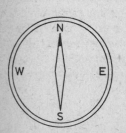

Fig. 1 Magnetic compass pointing north and south

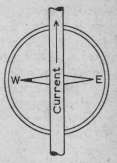

Fig. 2 Current deflecting a magnetic compass

soft iron loses practically all its magnetism when the current ceases. In the construction of electromagnets, soft-iron cores in the form of solid rod, iron wire or strips of thin iron sheet are employed. This magnetic effect is employed in electric bells, telephones, clocks and indicating devices. If a wire carrying a steady electric current is placed over a magnetic compass needle, as in Fig. 2, then with the current flowing in one direction the needle will be deflected one way, and on reversal of the current the direction of movement is reversed. This principle is employed in various types of instruments and meters used for measuring electric pressure, current and power, besides its application for power purposes to motors, generators, transformers and other devices. The discovery by Michael Faraday of electromagnetic induction, over a century ago, was the foundation of electrical power and machinery design. If electric conductors adjacent to one another carry excessive currents, then, in addition to the dangerous heating effect, there is a magnetic force between them which can cause movement and damage, but this does not occur in domestic installations with relatively low currents.

Chemical effects

If two wires from a small dry battery are dipped into a glass containing water, to which a few drops of vinegar are added, bubbles of gas will be seen to form on the wires when they are brought sufficiently close together *without* touching. This shows that the water has been split up into its constituent elements, oxygen and hydrogen, which are evolved as gases. If the same wires are stuck in a potato, about 2 mm apart, then the wire connected to the positive terminal of the battery will show a green discoloration due to the effect of the current on the chemical constituents of the potato. Do *not* try this with the mains supply as it is very dangerous. This effect is usefully employed in the electrodeposition of metals and in refining. It is also evident as apparent corrosion when there is a leakage of electric current from a cable or conductor to adjacent metalwork or damp surfaces. This is frequently seen on top of a car battery. The chemical effects are only produced with direct current (d.c.).

Arc effect

If two carbon rods connected to a suitable supply are momentarily touched together and then separated, an intense spark called an 'arc' is formed. With sufficient electric pressure across the gap, the current does not cease but persists in the form of an arc. The temperature of the carbon rods is about 3500 degrees Celsius and an intense white light is emitted, mainly from the incandescent carbons. This effect is usefully employed for searchlights, photographic processes, electric furnaces and electric arc-welding. A similar effect can occur with a break in an electric cable, when the arc may persist between the broken ends or to any adjacent metal-work, constituting a fire risk. Arcs also occur at switch contacts and wherever the current is interrupted by breaking a circuit in the wrong way instead of using a switch, such as disconnecting a plug and socket with direct current or with certain types of inductive circuits using alternating current.

Physiological effect
Electric shock to the various parts of the body is dangerous and may even be fatal. For this reason, care should be exercised with electrical connections. Switches should be 'off' or the fuses withdrawn when any alterations or repairs are being made to the wiring of a house. 'Safety first' must be considered at all times with electrical apparatus. An application of this effect is its use for electrocution in the course of justice in some states in America. Chapter 12 deals more fully with these aspects.

Electrical units

Before we are able to compare the sizes of different objects we must have some system of suitable units. A yardstick is suitable for measuring a length of cloth, but it is of no use in weighing coal or finding the hours of sunlight on a certain day. The scientific basic units include the *metre* (m) for length, the *kilogram* (kg) for mass and the *second* (s) for time. All the practical electrical units are related to these dimensions, but we are only concerned with those units in everyday use.

Electricity is a form of energy, which may be defined as the ability to do work; so, when energy is expended, the work is done. Electrical energy is charged for as units in the electricity bill, and the account may include a fixed charge and some charges for other services, such as the hire of an electric cooker.

The legal unit for the sale of electrical energy is the *kilowatt-hour* (abbreviated to kWh) and is usually referred to simply as the 'unit' of electricity.

Energy is equal to the rate of doing work multiplied by time, and the rate of doing work is called *power*. In everyday speech a motorvan may be said to be more 'powerful' for shifting a load than a horse and cart, meaning that the former does more work in a given time than the latter. If a labourer carries 10 kg of sand 10 m up a ladder, then 981 joules of energy have been expended, whether he takes 2 minutes or 2 hours over it; but in the former case the rate of doing work is sixty times faster than in the latter, and the power is therefore sixty times greater.

The electrical unit of power is the *watt* (symbol W). It is used when we say that a 60-watt lamp is more powerful than a 25-watt lamp. The unit for a quantity of electricity is the *coulomb* (symbol C), analogous to the litre as a quantity of water; but we are only concerned with the rate of flow of electricity, or current. This is the quantity of electricity flowing in a circuit in a unit of time, the second, and it is called the *ampère*, often abbreviated to amp or A. In a water main we are not concerned with how many litres it will hold but with the delivery in litres per second; so, in just the same way, we are interested only in the flow of current in an electric cable or how much current we must make flow in an electric fire to make it red-hot. A level water main full of water would not deliver any water to the domestic kitchen tap without the pressure behind the water; thus pressure effects the rate of flow. This is obtained either by the height of a reservoir above the main or by the pumps at the waterworks. In the electricity main or cable it is the pressure generated at the power station that causes the electric current to circulate around the supply system. This pressure is measured in *volts* (abbreviation V) and is present all the time, whether we have the electricity turned on, by means of a switch, or not. Thus the electrical power available depends on both the pressure and the intensity of flow. Hence *electrical power in watts is the product of the pressure in volts and the current in ampères*, or, with letter symbols, $W = V \times I$.

The difference in pressure, or *potential difference* (abbreviated to p.d.), across the ends of a circuit causes a current to flow if the circuit is complete. But what settles the magnitude of the current? If we have a new, clean water main, very little pressure will send the water along it, as the friction between the moving water and the walls of the pipe is very low. Suppose the main is half full of sand and gravel, then a very much higher pressure will be needed to deliver the same quantity of water in a given time; also, a smaller clean pipe would require a greater pressure and velocity to deliver water at the same rate. The total resistance to flow will depend on the length of the pipe, its internal cross-sectional area and its material wetted surface. In a similar manner, the conductor material, such as a copper, of an electrical circuit is like the bore of a water main and offers a resistance to the rate at which electricity flows through it. So it can be said that the rate of flow of electricity increases with the pressure and decreases with the resistance. This statement is the basis of Ohm's Law, which can be stated as: *Current is proportional to the voltage and inversely proportional to the resistance.*

The unit of resistance is the ohm (symbol Ω). A p.d. of 1 volt applied to a resistance of 1 ohm will cause a current of 1 ampère to flow; alternatively, volts divided by ohms gives ampères. With symbols, I stands for current, V for voltage and R for resistance; then:

$$I = \frac{V}{R}, \quad \text{or} \quad V = I \times R, \quad \text{or} \quad R = \frac{V}{I}.$$

If any two of these quantities are known, the third one can be found.

Power, in watts, is the product of voltage and current, therefore:

$$P = V \times I.$$

This product is true power only in direct current circuits and has to be multiplied by another quantity, called the 'power factor', to give true power in alternating current circuits. But in household installations this refinement is generally unnecessary. The watt is too small a unit for many installations, so the term *kilowatt* is employed, which is 1000 watts, and 1 kilowatt for 1 hour is 1 unit of energy.

Example 1. An electric fire, having a resistance of 40 Ω, is connected to a 200 V circuit. What current will it take, and what will be the power in kilowatts? If the cost of power is 1p per unit, how much does it cost to use this fire for 3 hours?

Current in amps $= \dfrac{\text{volts}}{\text{ohms}} = \dfrac{200}{40} = 5$ A.

Power in watts $\;\;= \text{volts} \times \text{amps} = 200 \times 5 = 1000$ W
$\qquad\qquad\qquad = 1\text{kW}.$

Energy $= \text{Power} \times \text{Time} = 1\text{ kW} \times 3\text{ h} = 3$ kWh.

If 1 kWh is 1 unit costing 1p, then cost $= 3 \times 1\text{p} = 3\text{p}.$

The old mechanical unit of power was the horsepower, which is equivalent to 746 watts, but under the new metric system motor ratings are stated in watts for machines that are driven by electric motors.

In the case of alternating current (a.c.) circuits and a.c. apparatus windings, the values of current and voltage are not coincident (in phase) due to inductive effects, which make the current lag behind the voltage. Capacity effects make the current lead the voltage. Therefore, the product of current and voltage values will not give the true power. In a.c. calculations this product is called *voltampères*, not watts, and the two different values of voltampères and true power watts are related by the *power factor*, which is the ratio

$$\frac{\text{Watts}}{\text{Voltampères}} \quad \text{or} \quad \frac{W}{VI}.$$

With a load of pure resistance the power factor is unity, or it may be expressed as 100%; but with inductive apparatus it is less than one, or below 100%. Low power factor means that more current is required for a given power, and as this adversely affects the supply system it is often penalised in tariffs.

Current required by motors

The output at the motor shaft is given in watts. Due to the motor losses, or power absorbed by the working parts in friction, windage and resistance, and converted to heat, the watts input at the motor terminals must be greater than the output, which is a measure of the *efficiency*.

For any machine, Efficiency $= \dfrac{\text{Work output}}{\text{Work input}}.$

For a d.c. motor, Efficiency $= \dfrac{\text{Watts output}}{\text{Volts} \times \text{Current input}}$

and Current input, in ampères $= \dfrac{\text{Watts}}{\text{Volts} \times \text{Efficiency}}.$

For small d.c. motors with an efficiency of about 75%, the current required from the supply is given by:

$$\frac{\text{Watts} \times 100}{\text{Volts} \times 75}.$$

With a.c. motors, the losses in addition to power factor still further increase the input current. With small motors, the power factor varies from under 70% to over 80%. Thus, for an a.c. motor:

Current input, in ampères

$$= \frac{\text{Watts output}}{\text{Volts} \times \text{Efficiency} \times \text{Power-factor}}.$$

Taking an average value of 75% for both efficiency and power factor, the current taken from the supply is given by:

$$\frac{\text{Watts} \times 100 \times 100}{\text{Volts} \times 75 \times 75}, \quad \text{or approx.} \quad \frac{1 \cdot 8 \text{ W}}{V}.$$

Measuring instruments

The pressure is measured by a voltmeter, which is *connected across* the mains or the apparatus concerned (in parallel). This is like a pressure gauge on a boiler which does not measure the steam consumption; in a similar way, the voltage is not a measure of the current used.

The current is measured by an ampèremeter, usually

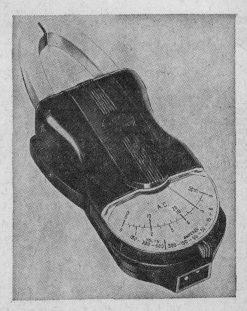

Fig. 3 A.c. clip-on volt/ammeter

(*Sangamo Weston Ltd.*)

called an ammeter, and is *connected in* the circuit, like a gas meter in the gas main (in series).

The correct method of connecting a voltmeter and an ammeter is shown in Fig. 4.

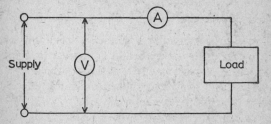

Fig. 4   Correct method of connecting a voltmeter and an ammeter

It is possible to connect the voltmeter across the supply mains, as it has a high resistance and thus will not take much current.

An instrument commonly used for checking alternating current (a.c.) is a 'Tong-Test' (Crompton) ammeter or 'Clipper' (Weston) volt/ammeter. The latter instrument is shown in Fig. 3, and this also measures voltage by using test leads which are plugged into one end of the instrument. A trigger opens the jaws to enclose a cable, and the current is measured by induction or transformer action and indicated on the selected scale; this is not possible with direct current (d.c.).

Example 2. A voltmeter that will read up to 250 V has a resistance of 10 000 Ω. What current will it take when connected across a 200 V supply?

$$\text{Current in amps} = \frac{\text{volts}}{\text{ohms}} = \frac{200}{10\ 000} = \frac{1}{50}, \quad \text{or} \quad 0 \cdot 02 \text{ A.}$$

The ammeter, on the other hand, must *never* be connected across the supply, whether it is from the mains or a battery, as it would take a very large current, due to its low resistance, and suffer internal damage.

Example 3. An ammeter that reads up to 20 A has a resistance of $\frac{1}{100}$ Ω. What current will it take if accidentally connected across (a) a 12 V car battery and (b) the 230 V mains? Theoretical values, excluding battery resistance, would be:

Fig. 5 Portable wattmeter

(Crompton Parkinson Ltd.)

(a) Current $= \dfrac{12}{0\cdot01} = 12 \times 100 = 1200$ A.

(b) Current $= \dfrac{230}{0\cdot01} = 230 \times 100 = 23\,000$ A.

In such a case the circuit fuses would cut off the supply, but the ammeter would also be damaged.

Example 4. What voltage is required across the same ammeter to give the full deflection of 20 A?

By Ohm's Law, Volts = Current × Resistance
$$= 20 \times \tfrac{1}{100} = \tfrac{1}{5}, \quad \text{or} \quad 0\cdot2 \text{ V.}$$

On d.c. systems the product of voltage and current can be measured by a wattmeter, but an ammeter and voltmeter can be used to serve the same purpose.

An a.c. wattmeter is a similar instrument but with the movement designed for alternating current; Fig. 5 illustrates a portable wattmeter for single-phase a.c. or d.c. reading up to 12 W, but these instruments are made for numerous ranges, in kW also.

Fig. 6 illustrates another portable testing set in common use. It has a selection of several ranges of volts and ampères, including milliamps, as well as resistance, measured in ohms, for d.c. and a.c. measurements. Some similar multimeter instruments also include scales of capacity, induction and decibels. This type of instrument is very compact and depends on electronic internal circuitry for its operation.

Supply meters

These meters are the property of the supply undertaking and are sealed on installation so that they cannot be tampered with by unauthorised persons. They combine the measurement of volts, ampères and time, and depend for their action on the magnetic effect, although on some direct current systems an electrolytic type which utilises the chemical effect of a current has been employed in the past. Instead of a

Fig. 6 Portable a.c. testing set

(Avo Ltd.)

pointer, as in an indicating instrument, the common single-phase meter has a moving element that rotates and drives a train of wheels, which records the number of units consumed on a number of small dials. The rotating disc runs between the poles of an electromagnet, and the speed of the disc varies with the current passing through the magnet coils due to eddy currents induced in the disc and producing a motor effect.

How to read an electricity meter

A meter is easy to read and gives the consumption of electricity. On the smaller sizes there are four black dials which register thousands, hundreds, tens and single units. The

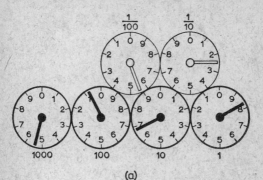

(a)

(b)

Fig. 7 Electricity meter dials

    (a)   10 A meter reading 5068 units
    (b)   50 A meter reading 67 029 units

hands of adjacent dials revolve in opposite directions. The red dials, which register in tenths and hundredths of a unit, may be disregarded since they are provided for testing purposes.

To read the meter start at the right-hand dial of single units. When the hand is between two figures, write down the *lower* figure; if between 0 and 9, always write down 9. Repeat the process with the other dials, writing down the figures in the order right to left. If the hand is *on* a figure (say 6), write down 5, not 6, unless the hand on the previous right-hand dial is between 0 and 1. On larger meters there are five black dials going up to ten-thousands.

Example 5. In Fig. 7 (a) is shown the dial arrangement of a 10 A meter; the reading is 5068 units. Fig. 7 (b) shows a 50 A meter; the reading is 67 029 units. Now try these simple readings from Figs. 8 (a) and 8 (b). What do these dials read? The answers are given at the end of the chapter, but don't look at them until you have tried your skill.

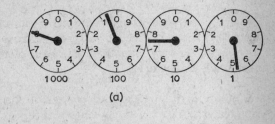

(a)

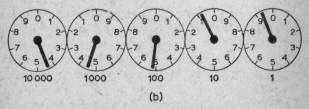

(b)

Fig. 8 Electricity meters to read

Charges for electricity

There are many scales of charges in different parts of the country and some perplexity has been caused by the variety of methods employed. It is realised in the Industry that rationalisation is desirable, and considerable progress in this has been made by the Area Boards.

Unfortunately, electricity cannot be generated at a steady rate and stored during the periods of low demand, like gas. Except for the relatively small amount that can be stored in 'secondary cells' similar to the 'accumulator batteries' used in motor cars, the power stations have to be able to supply whatever amount is called for 'on demand'. This means that the generating plant installed must be large enough to supply the 'peak' load, not just the average load on the system. The advent of the National Grid has helped to level out these peaks by the interconnection of stations, but they still exist. Weekday and Sunday load curves are shown in Fig. 9 for a large power station, where these peaks are evident despite the levelling effort obtained by interconnection.

The capital charges on the plant, together with the other standing charges, must be met, as well as the running costs of generation, which vary with the load. Four methods of framing tariffs are common:

1. *Flat-rate charges*. A high price per unit is charged for lighting, and lower prices for heating, power and other purposes.
2. *Block-rate tariff*. This tariff is generally available with a high price per unit for consumption up to a certain maximum number of units per month or quarter (first block), with much reduced prices per unit for the succeeding block or blocks. Meter rent is often an additional small item.
3. *Two-part tariff*. This consists of a fixed charge and a low price per unit. The fixed charge is arrived at in a variety of ways and may depend on the floor area of the premises, the rating valuation or the power of lamps installed. This amount is justified by the standing charges referred to earlier, as the cost and size of the generating plant are

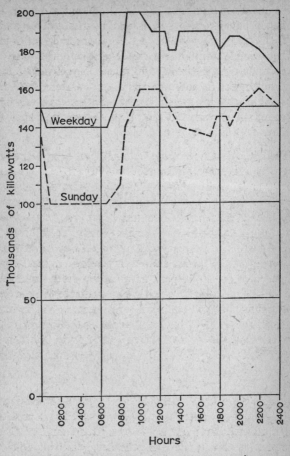

Fig. 9 Load diagram of a large power station

dependent on the size of the installations supplied with electricity.

4. *White meter tariff*. This tariff is designed to enable consumers to take advantage of off-peak rates at night, which is about half the normal daytime rate, without having

to install separate installations for off-peak metering. The meters have two sets of dials, and, by means of a separate time switch, one registers the night consumption, usually between 11 p.m. and 7 a.m., and the other registers the day consumption; the two quantities are charged at the appropriate rates. In addition, there are quarterly fixed and standing charges, as with the two- or three-part tariffs, to cover the unrestricted use of all equipment during the day, if required.

The flat rate is more usual for small flats or houses where electricity is used mainly for lighting. With additional heating and cooking loads, the block or two-part tariff is generally more economical for the householder, ensuring a contribution to the fixed charges of the supply authority and a reducing average cost per unit with higher consumption.

The white meter tariff favours the off-peak use of electricity for storage space heaters and water heating—in fact, any or all purposes during off-peak hours at low rates—whereas a separate off-peak tariff, which is also available without fixed or standing charges, requires separate meters, time switches and controls, and wiring installation (as distinct from the normal installation) which cannot be used during the daytime peak hours, and is intended for storage heating installations only.

Example 6. A small householder has the choice of paying for electrical energy *either* (a) at a flat rate of 2·5p per unit and 15p a quarter meter rent *or* (b) at a fixed charge of £2 per quarter plus 1p per unit for all purposes. Which would be the cheaper tariff if the average quarterly consumption is (i) 100 units and (ii) 200 units? At what number of units per quarter would it be worth while changing from one scale to the other?

(i) 100 units per quarter:
   Scale (a): 15p meter rent $+(100 \times 2 \cdot 5p) = 265p = £2.65$.
   Scale (b): 200p fixed charge $+(100 \times 1p) = 300p = £3$.

(ii) 200 units per quarter:

Scale (a): 15p meter rent $+(200 \times 2 \cdot 5p) = 515p = £5.15$.

Scale (b): 200p fixed charge $+(200 \times 1p) = 400p = £4$.

Scale (a) is the cheaper for 100 units, while scale (b) is the cheaper for 200 units per quarter.

The answer to the latter part of the question is most easily illustrated by a graph. A graph is a mathematical picture showing how variable quantities are related. Two axes are drawn on squared paper like two sides of a square and marked off to scale in the units of the related quantities. The above figures can be used and rearranged in the tables below, as illustrated in the graph given in Fig. 10.

Scale (a)	Units per quarter	0	100	200
	Total cost (pence)	15	265	515

The 'points' shown on the graph are joined by a straight line and titled 'Flat rate'.

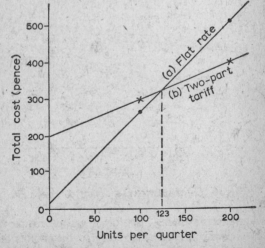

Fig. 10 Graphical comparison of electricity charges

Scale (b)

Units per quarter	0	100	200
Total cost (pence)	200	300	400

The 'crosses' shown on the graph are joined by a straight line and titled 'Two-part tariff'.

At least three sets of points should be taken to straddle the estimated consumption of the premises concerned. Where these two graphs cross one another shows the average number of units per quarter at which both tariffs are equal in cost and, on either side, which tariff offers lowest cost for any given consumption. This can also be solved by a simple calculation, as the total charges are equal at the intersection.

If x be the number of units, then:

$$15 + 2 \cdot 5x = 200 + 1x$$
$$\therefore \ 1 \cdot 5x = 185$$
$$x = 123 \text{ units, as obtained from the graph.}$$

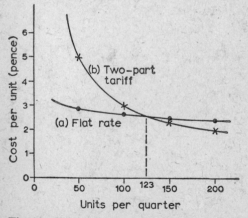

Fig. 11 Graphical comparison of cost per unit

As a matter of interest, two other graphs are plotted in Fig. 11, showing the cost per unit against the number of units per quarter, and these two curves also intersect at 123 units per quarter. These graphs bring out the advantage of the two-part tariff, as the electrical load increases with the use of more domestic appliances.

The block-rate tariff is very similar in effect to the two-part-tariff, and a third curve can be plotted in the same way, the only difference being that the curve will start from zero, rise steeply for the first block of units, less steeply for the next block of units and so on. The advantage it has over the two-part tariff is that the cost is much lower for periods of abnormally low consumption, because it does not have a fixed charge regardless of consumption.

Answers: Fig. 8 (a) 8075 units; Fig. 8 (b) 44 509 units.

2 Conductors, Insulators and Circuits

In two other public utilities, gas and water, the supply is distributed by the main pipes. To transmit electrical energy to the consumer cables of various types and sizes are employed consisting of central conductors, analogous to the bore of a water main, and the surrounding insulation, which can be likened to the wall of the pipe (see Fig. 12). Any material that allows an easy passage to electricity is called a conductor. All metals are conductors of electricity, but some are better than others. Conductors are required to provide an easy path for the current and should not have a high resistance, otherwise both drop of pressure and power loss will occur which will cause unwanted heating. Low-resistance conductors are

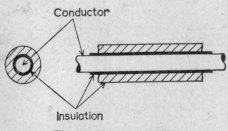

Fig. 12 Single-core cable

therefore required to carry current to the points of utilisation and control, such as the wiring connecting the switches and lamps to the supply The most common metal for wires and cables is copper, which is specially refined to have high conductivity. Copper is easily tinned and soldered, and is also a ductile metal.

Aluminium is also used, particularly for large sections, to carry heavy currents, but it cannot easily be soldered and a special technique has to be used for jointing.

Brass is employed for contacts and terminals, e.g. the brass plungers inside a lampholder.

Carbon is a non-metallic conductor and is used for sliding contacts on brass or copper, e.g. the carbon brushes that lead the current into the revolving armature of a motor. The lead of a pencil is a conductor, and it is possible to get a shock from a lead pencil if it is poked into an electric socket.

Iron is a poorer conductor than copper, due to its higher resistivity, and is seldom employed as its magnetic properties may also cause extra losses.

Factors affecting the resistance of a conductor

The greater the length of a conductor, the higher will be its resistance, and for a given length and material the resistance can be lessened by increasing the area of cross-section. If l is the length of a conductor, a its cross-sectional area and ρ its resistivity, which depends upon the material of the conductor, then resistance is given by

$$R = \frac{\rho \times l}{a} \text{ ohms.}$$

The 'resistivity' of any material is the resistance between the opposite faces of a cube of the conductor which has an edge of unit length, the value of resistivity depending on the unit used (see Fig. 13). Since copper has high conductivity it has low resistivity, and the resistance between the opposite faces of the cube is very small. Such very small numbers less than one can be written as a decimal with a number of noughts in front of the first significant figure, but for convenience we denote thousandths by 'milli' and millionths by 'micro', placed in front of the unit name. Thus 5 milliampères is five thousandths of an ampère, or $\frac{5}{1000}$ ampère, and 2 microhms is two millionths of an ohm, or $\frac{2}{1000000}$ ohm. For copper the resistivity, ρ, is approximately 0·0172 ohm per square milli-

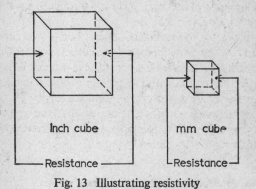

Fig. 13   Illustrating resistivity

metre per metre. As metals get hotter their resistance increases, though the opposite effect occurs with carbon; but the above value for copper is sufficiently accurate for our purpose.

Example 7. A copper wire has a diameter of 1·78 mm and a cross-sectional area of 2·5 mm². If its resistivity is 0·0172 in metric units (mΩ), what will be the resistance of a length of 1 km?

$$R = \frac{\rho \times l}{a} = \frac{0·0172 \times 1000}{2·5} = 6·9 \; \Omega \text{ approx.}$$

(Convert the length in kilometres to metres and note that all the units must be of the same kind.) This size of wire is used in cables, and 1 km of single strand will weigh about 22·2 kg without the covering of insulation. As approximate figures for calculating weights, copper weighs about 8·9 g/cm³ and about a third more than iron.

Example 8. If the above copper wire of 2·5 mm² were cut in half and run side by side, what would now be the resistance of the two wires connected together in parallel?

The length is now 500 m, but the cross-sectional area is doubled to 5 mm², so by proportion:

$$R = 6·9 \times \frac{500}{1000} \times \frac{2·5}{5} = 1·73 \ \Omega.$$

Insulators

Electricity will escape to exposed metalwork and earth if not prevented, and in so doing will become dangerous. To keep it to its proper path in the conductor it is necessary to insulate the conductor. Insulators are necessary to separate the conductor from other conductors, exposed metalwork and earth. Insulators are substances that are poor conductors of electricity and serve to prevent, as far as possible, an electric current straying from its conductor path. The pressure or voltage on an electric cable, while sending the current along the conductor, also tends to force a very small current through the wall of insulation around the conductor. If the insulation is insufficient or faulty, there will be a leakage current, which will flow to the return conductor (short-circuit) or to earth and return to the power station by any easy path, such as the water mains. Due to the contents of the soil, chemical action will occur, causing corrosion, but this is more prevalent with direct current than with alternating current systems. The insulation can be likened to the wall of a water pipe, which will only stand a certain hydraulic pressure before it begins to leak and may eventually burst. In a similar manner, if the pressure or voltage on an insulated wire is increased beyond a safe limit, the insulation will break down. Electric wires and cables are insulated for definite working pressures and should not be used for higher voltages. Thus a length of bell wire is amply insulated for the 3 V pressure obtained from two dry batteries, but it is most unsafe to use such wire for connecting an appliance to 240 V mains.

The resistivity of insulators is very high, and instead of writing a very large number with many noughts behind it we denote thousands by 'kilo' and millions by 'mega'. Thus 2000 watts is called 2 kilowatts, and 600 000 000 ohms is called 600 megohms ($M\Omega$). The insulation resistance of cables is measured in megohms and decreases with increase of length, as there is a larger area of insulation—of the same thickness —for the leakage current to pass through.

Examples of insulators are india-rubber, both pure and vulcanised, polyvinyl chloride compound (p.v.c.), gutta-percha, ebonite, mica and slate. Some insulators are hygroscopic, i.e. absorb moisture, such as paper, cotton, silk, various artificial fibrous materials and asbestos. As a general guide, good heat insulators are also electrical insulators, so that they tend to retain any heat in the conductor generated by the passage of the electric current. Continual overheating will cause some insulators, such as rubber compounds, to become hard and brittle, and will soften others, such as p.v.c., so cool operation is essential for safety.

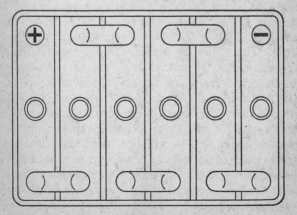

Fig. 14 Car battery, six cells in series

Simple circuits

The two simple arrangements of conductors and connected apparatus in a circuit are called 'series' and 'parallel'. Two horses harnessed to a trap tandem may be said to be in series, while two horses pulling a dray side by side would be in parallel. The six cells of a 12 V car battery are connected in series as shown in Fig. 14. Since each cell gives 2 V, the total pressure adds up to 12 V, and the current is common through each cell. If 8 A are taken for lamps, then this amount of current flows through each cell. Two dry cells for an electric bell each give 1·5 V. In Fig. 15 they are shown connected in

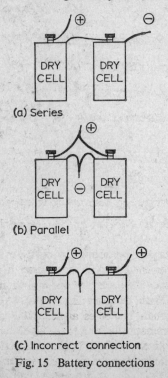

(a) Series

(b) Parallel

(c) Incorrect connection

Fig. 15 Battery connections

series at (a), when the total voltage is 3. In (b) the similar terminals are connected together, when the pressure available is only that of one cell, namely 1·5 V; but each cell will supply half the current, i.e. the total current in a parallel circuit is the sum of all the currents. In (c) the two negative wires are joined together, and the two positive terminals are shown going to the external circuit. This is incorrect, as the voltage of one cell is trying to send the current one way round the

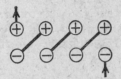

(a) Series, maximum voltage

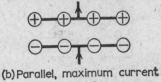

(b) Parallel, maximum current

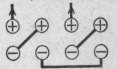

(c) Wrong connection, no current

Fig. 16 Series and parallel connections

circuit from its positive terminal while the other cell is trying to send the current round in the reverse direction. The net result is that the voltages are in opposition and no current will be supplied. With batteries there must be a positive and negative terminal. The series and parallel connections are shown diagrammatically for a four-cell battery in Fig. 16 (a) and (b) respectively, while at (c) the connections are wrong.

Electrical energy from the mains is distributed with parallel

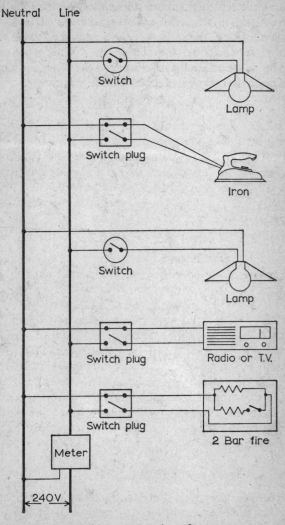

Fig. 17   Parallel connection of apparatus

connections, for which the voltage is maintained constant, and all the separate currents add up to give the total load current on the system. The same parallel method is employed inside the house, the various currents adding up to the total current supplied. But the outlet-point wiring, switch and apparatus are connected in series, since the switch must open and close the circuit, and all these items will carry the same current.

Example 9. Suppose that in one room of a house an electric iron is being used; the iron will take 2·1 A, and a 60 W lamp will require 0·25 A. In another room the family is listening to the radio or watching television in front of an electric fire with one bar switched on. The T.V. set will take 0·5 A, a 100 W lamp 0·42 A and the 1 kW fire 4·2 A, so the total current supplied to the premises is:

$$(2·1+0·25)+(0·5+0·42+4·2) = 7·47 \text{ A,}$$

say, 7·5 A in round figures.

The size of wire for each piece of apparatus must be large enough to carry the current and must also be properly insulated to withstand the common pressure of 240 V. The switches are in series with the individual appliances they control, as shown in Fig. 17. From this it will be seen that each piece of apparatus has a common voltage of 240 and that the various streams of current combine to give a total of 7·5 A through the electricity meter. Note the switch on the electric fire for the second element; the switch plug will disconnect the fire completely. The radio or T.V. set will also have its own switch, but the pressure is on the connecting lead as long as the switch of the switch plug is closed.

Fuses and circuit protection

The heating effect of an electric current may become dangerous, but this effect can be used to open a circuit in which the current is excessive. Fuses are provided to protect both the wiring and connected apparatus, and in so doing protect the adjacent structure of the building against fire. Fuses are sometimes called 'cut-outs' because they cut out the defective circuits. The term fuseboard is given to the complete assembly of a case containing a group of fuses and contacts, but the term distribution board is more general, as it also covers groups of automatic switches that disconnect the circuits when over-loads occur. Such miniature circuit-breakers are frequently employed in modern installations, but fuses are more common.

Fuses consist of short lengths of a suitable wire that will safely carry the normal current of the circuit, but with any sustained overload the wire melts. A very large current will blow the fuse immediately, while a lesser current may take several minutes to melt the wire.

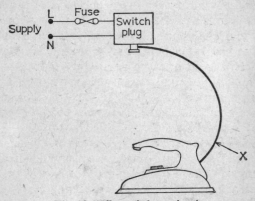

Fig. 18   Effect of short-circuit

For simplicity no fuses were included in Fig. 17, but the various lighting and heating circuits would be divided up and suitably protected by fuses. The electric iron would be supplied through a flexible lead from the switch plug, and a fuse would be inserted between the connecting wires and the supply terminals, as shown in Fig. 18. If everything is in order, the current will not cause any heating of the fuses and current will be supplied to the iron. Now, suppose that the two wires of the flexible lead are worn and that so much of the insulation is rubbed off that the two bare wires can come into contact at the point X. This would cause a 'short-circuit', and the current would take a short or easy path of very low resistance at the fault instead of going through the heater element of the iron. The current would surge up to a very large value and the circuit fuse would melt. The current would be automatically cut off, before any damage was done to the rest of the circuit, as it could not get across the gap left by the melted fuse.

It is essential that the correct fuse should blow when a fault occurs and that a fuse should not blow unnecessarily. For this reason, the sizes of fuses are graded to suit the various circuits, such as 5 A for the lighting circuit and 20 A for heating circuits. The main fuses should not melt and cut off the whole supply to a house when a fault occurs on a sub-circuit; this points to bad installation design or wrong choice of fuse. Therefore, it is important to replace 'blown' fuses with the correct size of fuse wire or fuse cartridge, which should have the current-carrying capacity of the smallest cable or flexible cord in the circuit it protects. The sizes of fuse wire for re-wireable-type fuses are given in Table 1.

If a fuse blows repeatedly, it is wrong and dangerous to replace the fusewire with a larger size than it should have or with any odd wire that happens to be at hand. The proper course is to investigate the cause of the fusing and remedy the fault. Cartridge fuses are safer and operate more quickly than rewireable fuses, but they are a little more expensive. The miniature circuit-breaker is the best form of circuit protection, and the extra expense is generally worth while, as it saves trouble and mistakes in replacing fuses and costs nothing to reset when it has operated. It is commonly used in America and is gradually becoming more popular in this country.

An important fuse that is too often overlooked is the small

cartridge fuse in 13 A plugs and fused adaptors. This should always be checked when connecting appliance flexibles to ensure that only 3 A fuses are inserted for the smaller appli-

Current rating of fuse A	Nominal diameter of wire mm
3	0·15
5	0·20
10	0·35
15	0·50
20	0·60
25	0·75
30	0·85
45	1·25
60	1·53

Table 1 Fuse elements for semi-enclosed fuses
(plain or tinned copper wire)

ances. (See also 'coarse' and 'close' excess current protection in Chapter 4.)

Need for switches in live side

The incoming supply cable to a house consists of the live or line wire and the neutral wire. Switches make and break electrical contacts, thus closing or opening an electric circuit. It has already been explained why fuses are used, and now Figs. 19 and 20 show alternative connections of a lamp and

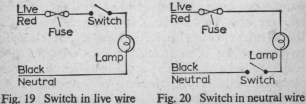

Fig. 19 Switch in live wire   Fig. 20 Switch in neutral wire
(wrong)

switch. In each case the switch will perform its function, but Fig. 19 is the *correct method* and Fig. 20 is *wrong*. The invariable rule must be for all switches to be in the *live* wire, which is run with a *red* wire. The reason is that the neutral (*black*) wire is connected to earth at the power station, while the live (*red*) wire is connected to the supply voltage, which may be as high as 250 V above earth potential. Thus in Fig. 19 the supply voltage *above* earth is disconnected from the lamp when the switch is open and it is safe. With the switch as shown in Fig. 20 the lamp is still connected to the full voltage, which is dangerous, despite the switch being open, if a person replaces a broken lamp thinking the lampholder to be 'dead'.

Earthed neutral

The maximum allowable single-phase pressure for domestic consumers is 250 V, and this is the voltage between any one line conductor and the neutral. The zero datum of the neutral is fixed by an earthed neutral point maintained by the supply authority. On alternating current systems the supply authority may provide a multiple-earthed neutral system, which means that the neutral conductor is permanently earthed at several points. Where one pole of a two-wire supply is connected to earth, this will affect the installation in a house since all fusing must be single-pole, i.e. single-pole m.c.b. or fuse assemblies are employed and single-pole fuseboards with neutral bar terminals for solid connection of the neutral circuit cables. There must be no break in the neutral conductor of the installation throughout the house. The earthing of the neutral is a safety measure that effectively prevents any part of the installation (exposed metalwork) becoming 'live' at a greater voltage than the single-phase voltage to earth in the event of a fault. Where neither pole of a two-wire supply is connected to earth, double-pole fuseboards must be used for both main and sub-circuits.

Medium and low voltages

For economic reasons, the supply authority distributes electricity at as high a voltage as possible, and street mains are usually at a pressure of 415 V between phase lines in a three-phase, four-wire a.c. system, or they may be at 480 V between 'outers' in a three-wire d.c. system. In both cases the neutral wire is generally at earth potential, and this limits the voltage between any line conductor and earth to 240 V. Thus the pressure of any live wire above earth is limited as a safety precaution. Up to 50 V is called extra-low, up to 250 V is low and up to 600 V is medium voltage.

Direct current and alternating current

A direct current is one that flows in one direction only, which is said to be from the positive (+) terminal or pole to the negative (−) terminal or pole.

These markings will be seen on accumulators and some direct current instruments, which will only indicate when the positive wire is connected to the positive terminal and the negative wire to the negative terminal.

An alternating current does not flow constantly in one direction but reverses in a periodic manner. Over one half of the cycle the direction is positive and during the next half-cycle it is negative. The number of complete alternations per

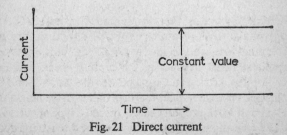

Fig. 21   Direct current

second is called the *frequency*, denoted by '*f*' or '∼', and the unit is the *hertz* (Hz). There is no right and wrong way of connecting the two wires to a.c. instruments, as there is with d.c. voltmeters and ammeters.

Direct current can be represented by a horizontal straight line in the graph shown in Fig. 21, while alternating current is depicted by the wavy curve of Fig. 22. An alternating current of 1 A can do just as much work or light a lamp just as brilliantly as a direct current of 1 A. On a d.c. system the voltage has a steady value, while the alternating voltage varies in the manner indicated but has just as much effective pressure to send the current around the circuit.

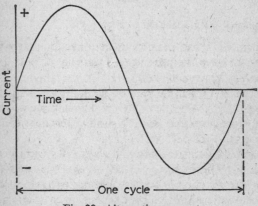

Fig. 22 Alternating current

Domestic apparatus that is marked 'Universal' for a certain voltage will operate equally well on either type of current; otherwise, it is specified as suitable for only one type. All appliances that depend on the heating effect, such as radiators, irons, kettles, etc., can be used on either a.c. or d.c. at their rated voltages. Apparatus depending on the magnetic effect will only work properly on the specified type of current, while in applications of the chemical effect d.c. is required.

One can get an equally painful electric shock from either a.c. or d.c., though the effects vary, but a.c. is considered to be the most dangerous.

Standard voltages and frequency

The term 'consumer's voltage' denotes the voltage at the incoming supply terminals, declared by the supplier. This corresponds to the term 'declared pressure' in the Electric Lighting Acts. Domestic consumers are supplied on low voltages and the standards for new systems and installations are:

Direct current systems (three-wire):
Consumer's voltage: 240 V and 480 V.

The former figure is that supplied for household use and is the voltage between the live conductor, which may be either positive or negative, and the neutral conductor. The latter figure is the voltage between the positive and negative outers (both 'live' conductors); it is used for large motors and large apparatus found in factories but not in the home.

Alternating current systems (three-phase):
Consumer's voltage (declared): between neutral wire and each of the phase conductors, 240 V; between any two phase conductors, 415 V.

The former value is used for domestic installations, while the latter figure is used for larger power purposes. Large buildings and institutions may have the incoming supply at the higher voltage, but it is divided into the lower voltage for the separate divisions and sub-circuits.

Frequency is now standardised in this country at 50 cycles per second, or 50 Hz. This means that there are fifty of the waves shown in Fig. 22 in every second, or the time of one cycle is 0·02 second. Apparatus of American origin is often only suitable for 60 c/s, which is commonly employed there, though there are also 25-cycle systems. The frequency does not affect apparatus depending on the heating effect, as will be evident from page 37, but small motors, transformers and any apparatus depending on the magnetic effect of a current will only operate satisfactorily at the correct frequency for which they are designed. The above figures apply to a large number of supply authorities, but there may be systems existing in which the voltages vary from 100 to 250 and the frequencies between 25 and 100 c/s. Any supply undertaking will give details of its standard voltages and conditions of supply, as well as the different types of tariffs available, together with any special local limitations on connected apparatus. These should be obtained for reference purposes for new installations.

Effects of incorrect voltage

It is essential that apparatus of the correct voltage is installed, since if the supply voltage is too high damage may occur, while if it is too low the performance will be unsatisfactory. The voltage supplies the necessary impetus to send the current around the circuit to do the work required. Thus, if a 100 V metal-filament lamp is put in a lampholder connected to a 200 V supply, twice the proper pressure is applied and twice the current will flow through the lamp (while it lasts). Now, power (in watts) is proportional to the voltage times the current, so in this case the power applied to the lamp is $2V \times 2I = 2^2VI$, i.e. four times normal, and, of course, the lamp burns out. If, on the other hand, a 200 V lamp is put on-to a 100 V supply, the current is only half that required, so the power is $(\frac{1}{2})^2$, i.e. a quarter of normal, and the lamp is very dull. Small differences of voltage have an effect on the illumination and the life of the lamp, both of which will vary considerably with the differences in commercial voltages. Thus, with a 1% increase of voltage on a metal-filament lamp, about $3\frac{1}{2}$% more light is obtained, but the useful life is shortened by some 12%. With a 1% drop in voltage, there is about $3\frac{1}{2}$% less light, but the useful life is lengthened by about 13%.

Incoming supply and methods of distribution

The provision of the supply is the responsibility of the supply authority which lays the service cable from the distributor running along the roadway. There may be a charge for opening up and connecting, and it is advisable to check that a large enough service cable is installed, though usually ample cable size is provided. The service cable is terminated at the supply authority's fuses, which are sealed so that they cannot be interfered with by unauthorised persons and are on the incoming side of the electricity meter. The supply then goes

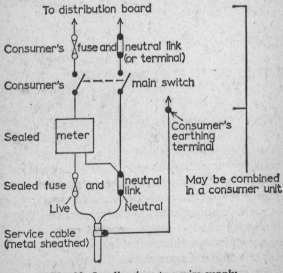

Fig. 23 Leading in a two-wire supply

to the main switch, which connects the installation to the meter, and thence through the consumer's fuse to the distribution board. The connection and maintenance of the meter is the responsibility of the supply authority.

This arrangement is illustrated in Fig. 23, which shows the leading in of a two-wire system. The service cables are connected to the distributors, which go back to the power station or sub-station. A three-wire system of d.c. distribution is shown in Fig. 24. The live outers are marked positive (+) and negative (−), and the earthed neutral (±) is shown

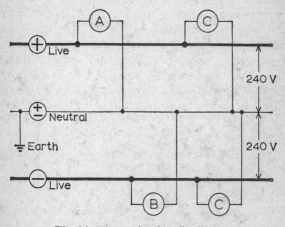

Fig. 24 Three-wire d.c. distribution

in between the two outers. Two-wire connections to separate domestic consumers are shown at A and B; in each case, one conductor brought into the house is connected to a live conductor and the other one to the earthed neutral. Connection C shows the connections to a large consumer, the wiring being divided into two entirely separate parts and the load being approximately balanced on each side.

Fig. 25 indicates diagrammatically a three-phase a.c. four-wire system, which is much more common. The three-

pointed star represents the transformer in the sub-station, the
midpoint of which is earthed and connected to the neutral
conductor. The three phase conductors form the live conductors, and are coloured red, yellow and blue. Inside the
house RED is used for the live wire or phase conductor and
BLACK for the neutral, the two different colourings being
used throughout in A, B and C. With a large building the
three-phase, four-wire supply may be brought in and divided
up as shown at D. Separate consumers having two-wire
connections are shown at A, B and C, the different houses

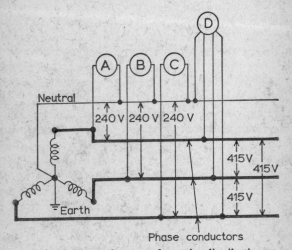

Fig. 25 Three-phase a.c. four-wire distribution

being balanced along the road between the three phases. A
large consumer might require to drive large motors; then
the connections would be as indicated at D, so that the line
voltage is 415 and the three voltages to neutral are each
240 V. In the diagrams the joints are shown by a black dot,
and where two wires cross over without any connection the
black dot is absent.

3 Switchgear, Distribution and Wiring

The main fuses protect the whole installation and may be housed in a separate case or included with the main switch.

Modern practice is to use either iron or moulded insulation cases for the main switch, which generally includes the main fuse(s) also. The main switch is of the double-pole type, so that the installation is completely disconnected from the supply when the switch is in the 'off' position. With large a.c. installations, a triple-pole switch is used to connect the three-phase supply to the distribution board, where a single-phase supply is given to separate pairs of cables, made up by one connection from each phase and the common neutral link.

The main fuses are contained in porcelain carriers and the fuse wire is completely protected from accidental contact. The cover of combined switch fuses is provided with a mechanical interlock, which prevents the cover being opened unless the operating handle is in the 'off' position. It is then safe to withdraw the fuses for replacement or to isolate the installation from the supply without the possibility of making or breaking the circuit on the fuse-carrier contacts, or of coming into personal contact with live parts. The iron case must be earthed, and a terminal is provided for this purpose. The earthing wire should not be less than 6 mm² copper cable. Standard sizes for main switches suitable for house installation range between 15 A and 100 A. A double-pole main switch and fuses is shown in Fig. 26, while Fig. 27 shows a triple-pole switch fuse with neutral link.

Splitter switches can be employed for small installations.

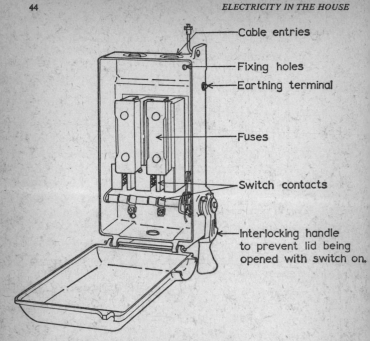

Fig. 26 250 V 30 A double-pole switch fuse
(Midland Electrical Manufacturing Co. Ltd.)

These combine the main switch and fuses for the separate circuits. Such an arrangement is cheap, but it is only adopted for small installations where there are very few sub-circuits. Splitter switches usually contain a main switch and two or three 15 A single- or double-pole fuses in one case for two-wire installations. There is no separate consumer's main fuse. A splitter switch with two single-pole and neutral ways is shown diagrammatically in Fig. 28, and illustrated in Fig. 29 (a) is a three-way, double-pole pattern. A three-way, single-pole and neutral splitter switch is shown in Fig. 29 (b).

A development of the splitter unit, known as the consumer's control unit, is commonly used in modern house

SWITCHGEAR, DISTRIBUTION AND WIRING   45

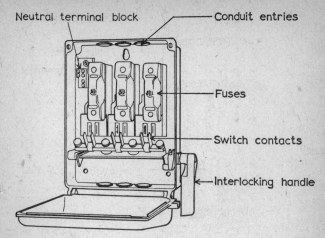

Neutral terminal block — Conduit entries

Fuses

Switch contacts

Interlocking handle

Fig. 27 500 V 30 A triple-pole switch fuse showing neutral terminal

(Bill Switchgear Ltd.)

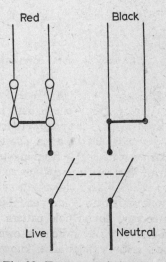

Red   Black

Live   Neutral

Fig. 28 Two-way switch splitter

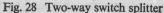

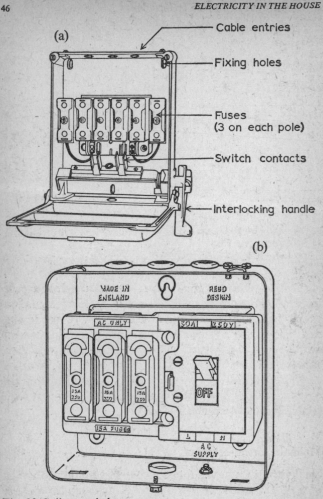

Fig. 29 Splitter switches

  (a) Three-way, double-pole, 250 V, 45 A *(Bill Switchgear Ltd.)*.
  (b) Three-way, 250 V, 37 A—single pole and neutral *(Midland Electrical Manufacturing Co. Ltd.)*

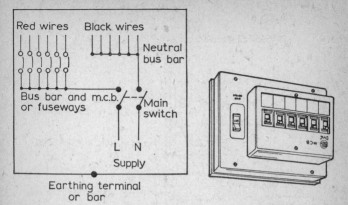

Fig. 30   A typical six-way consumer's control unit with miniature circuit-breakers

(Bill Switchgear Ltd.)

installations. This unit combines a 60 A or 100 A double-pole main switch with single-pole circuit fuses or miniature circuit-breakers (m.c.b.s) and a neutral terminal bar in one case. The consumer's main fuse is not necessary with such an arrangement, and the circuit fuses or m.c.b.s are designed to suit the usual house circuits, i.e. up to twelve ways of varying capacities between 5 A and 45 A for lighting, socket outlet, cooker and similar circuits. Fig. 30 shows a typical consumer's unit with m.c.b.-ways.

Miniature circuit-breakers

Small-capacity, automatic circuit-breakers for the protection of 5 A, 15 A, 20 A, 30 A and 45 A final sub-circuits are becoming popular in replacing fuses in the better domestic installations. They are grouped in distribution panels and consumer units in the same way as fuses, with similar or various ratings to suit the capacity required in the sub-circuits. The size is almost the same as a fuseway up to 30 A

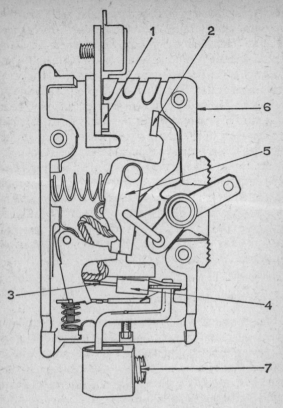

Fig. 31 Mechanism of miniature circuit-breaker
(Dorman & Smith Ltd.)

1. Silver tungsten contacts for long contact life
2. Wiping action keeps contacts clean
3. Thermal overload tripping
4. Magnetic short-circuit tripping
5. Quick break trip free mechanism
6. Universal flush and surface fixings
7. Reversible cable clamp connectors.
   Provision for individual locking

capacity, and some manufacturers make basic distribution units into which rewireable or cartridge fuse-carriers or miniature circuit-breakers can be plugged as required without altering the design of the distribution unit. The cost is only about 60p per way more than that of a fuseway, which is not a great deal extra in relation to the cost of an installation and can be justified by the advantages gained. These are elimination of fuse rewiring or replacement—to say nothing of replacement by the wrong size of wire or cartridge—visual indication of the opened circuit and the 'free handle' action, which prevents closing the circuit if the fault persists. The circuit-breakers are sealed for safety and thus are immune from interference. These miniature circuit-breakers are tested to break an a.c. short-circuit current of 3000 A, and give good protection and discrimination by thermal operation on overload, and a magnetic trip under heavy fault conditions. Such a breaker is shown in section in Fig. 31.

Main circuits and sub-circuits

The main switch and fuses look after the total current demanded by the premises. The total current can be compared to a river that gets its total flow from a number of tributaries, but the flow is reversed. The sub-circuits are the tributaries, and each one must be protected by its own fuse, so that if one section develops a fault it is cut out and isolated without the whole installation having to be disconnected.

In Fig. 32 six lamps are shown connected to one fuseway, in which L indicates a lamp and S a switch. This is a conventional diagram, and in practice 'looping out' is employed, as will be explained later on. It should be noted that the switches are connected to the RED WIRE on the live or line side, while the lamps are connected to the BLACK WIRE on the earthed, neutral side.

With larger installations the various circuits are supplied from a consumer control unit (Fig. 30) or a switch fuse and separate distribution board, the cables being suitably routed for convenience and economy. With large rooms two separate circuits may be used, so that if one circuit develops a fault there is enough lighting left for safety. Socket outlets rated at 13 A are given separate radial group circuits or are arranged on a ring circuit.

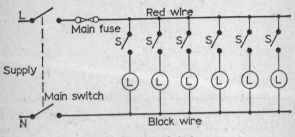

Fig. 32   Circuit of six lamps

If electrical energy is supplied through two meters, one for lighting and the other for power, then entirely separate main circuits must be run with no possibility of interconnecting the different types of load.

The main distribution board is supplied, from the main fuses to the bus-bars, with cables large enough to carry the total current and with some margin for future increase of load. Such an arrangement is shown in Fig. 33 for a distribution board with five ways. Each circuit has two fuses,* one connected to each bus-bar and supplies six lamps and their associated switches. This is the most convenient arrangement for medium-sized houses when all the current passes through one meter. With an 'all-in' tariff, one distribution board is sufficient, though lighting and power circuits are separated because of the different current capacities.

With separate scales of charges, two separate distribution boards are required and the installation may be more expen-

* For a two-wire, non-earthed installation in this case.

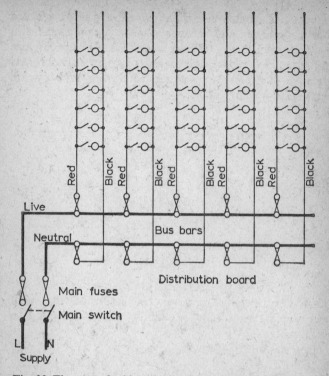

Live

Neutral   Bus bars

Distribution board

Main fuses

Main switch

L N
Supply

Fig. 33 Five-way, double-pole distribution board and circuits

sive. These arrangements are shown by single-line diagrams in Fig. 34. Water heaters and cookers are also run on separate circuits, sometimes with separate meters, main switches and fuses. Sub-distribution boards are necessary in large establishments with long cable runs, where the distance from the main distribution board would be too far for some final sub-circuits due to excessive voltage drop. A sub-board is fed from a pair of cables brought back to the main distribution board. If this sub-board is connected directly to the main switchboard, its own switch fuse gives protection to the cables; but if it is connected to a main distribution board, the correct practice is to connect it to a fuseway, as indicated in Fig. 35. Fig. 36 shows the wrong method and is bad practice, and the sub-main cables would have to be as large as the

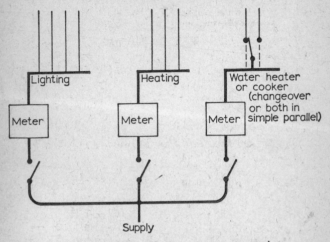

Fig. 34   Single-line diagram, separate metering

mains to the supply. In Fig. 36 the individual circuits from the sub-board are protected, but *not* the sub-main cables if they are smaller than the main cables, and as they add their load current to that already demanded by the main distribution board the main fuse(s) will not give sufficient protection if a fault occurs on the cables to the sub-board. Note the incorrect connection of three switches in this figure and the wrong provision of fuses in the neutral conductor of the installation, since the neutral is earthed.

For buildings with several floors of large extent, a suitable arrangement is shown in Fig. 37, in which each pair of rising cables is connected to a fuseway on the main distribution board. Fig. 38 is an example of bad practice: a single pair of rising cables is shown. The load on the cables decreases as

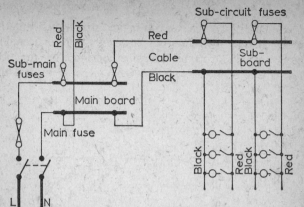

Fig. 35 Sub-board fed from fuseway on main board

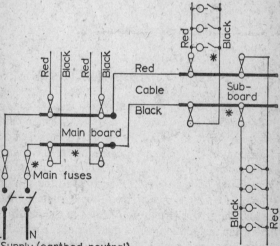

Fig. 36 Example of bad practice. Sub-board directly connected
to main board

Note other mistakes: Switches incorrect in black wire
and fuses should not be fitted in neutral.

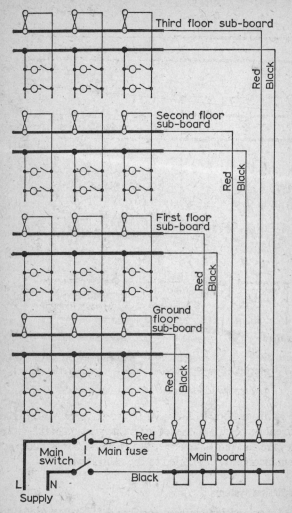

Fig. 37 Main distribution board with sub-boards connected by
         fuses

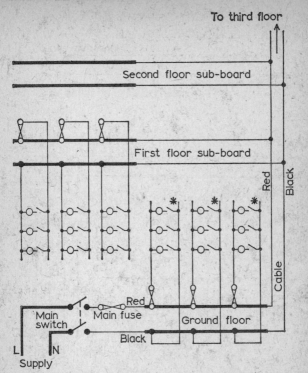

Fig. 38. Example of bad practice. Sub-boards on three floors tapped off riser. Rising cable not properly protected

Note: Switches incorrect in neutral.

they go from floor to floor, so the main fuses cannot give sufficient protection to the more remote portion of the cables if they are also reduced in size. The ground-floor switches are shown incorrectly connected to the neutral wire.

Installation systems

The various wiring systems employed for installations are generally designed for economy, length of life or mechanical strength, depending on the class of work required and the nature of the building. With any system, the avoidance of danger to life and the safety of the structure should take precedence over other considerations.

In Table 2 below various wiring methods are listed in order of merit, but the comparisons must not be taken too literally.

System	Relative			Labour required	Extensions and renewals	Protection against		
	Cost	Life	Time to install			Fire	Damp	Mechanical damage
1 Conduit (metallic) (plastic)	100 70	100 100	100 75	Skilled Skilled	Difficult Fair	Good Good	Good Good	Very good Good
2 Mineral insul. metal-sheathed	110	150	90	Skilled	Fairly easy	Very good	Good	Good
3 Metal-sheathed	75	90	85	Skilled	Fairly easy	Fair	Fair	Fair
4 P.v.c.–sheathed	60	90	50	Semi-skilled	Easy	Poor	Fair	Fair
5 Cleat (temporary)	45	50	50	Semi-skilled	Easy	Poor	Poor	Poor

Table 2 Wiring methods compared approximately

1 *Conduit systems*

These are the most satisfactory systems, because rewiring can easily be done, if the system is properly designed, without disturbing the building's structure and finishes. If rewiring is to be provided for, elbows and tees should not be used, since they prevent the easy drawing in of new cables when necessary. Steel tubes with screwed joints can be concealed in the walls during building, and for surface work they provide the best mechanised protection for cables; and by suitable precautions it is possible to make the wiring practically waterproof. Light- and heavy-gauge conduit is available, which may be solid-drawn seamless, seam-welded with almost imperceptible joint or 'close-joint', which is a longitudinal joint butting together. The heavy-gauge welded tube is used for high-class screwed work; light-gauge welded or light-gauge close-joint conduit is used for cheaper work. The latter should not be used when buried as it is not watertight, nor should it be employed for wiring for supplies over 250 V.

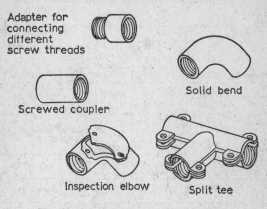

Adapter for connecting different screw threads

Screwed coupler

Solid bend

Inspection elbow

Split tee

Fig. 39   Some screwed conduit fittings

Steel conduit is generally finished in black enamel, inside and out, but for very damp situations galvanised or zinc-impregnated finish is used. Gas barrel is fitted for some outside work; it is thicker than conduit and needs adaptors to connect it to conduit. Gas barrel sizes are given in inside diameters, while conduit is specified by its outside diameter. Special fittings are necessary with conduit, consisting of elbows, tees, bends, junction boxes and inspection boxes; these may be screwed, or they may depend on grip for continuity, as illustrated in Figs. 39 and 40. The insulated wires are drawn into the conduit after it has been installed, but the tube must not be packed tight with the wires; there should always be room to draw in additional conductors of the size already inside. It is possible to accommodate a greater number of wires in a short straight run than when there are a number of bends and

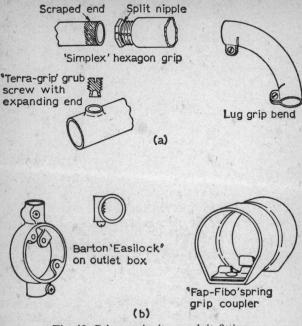

Fig. 40 Grip continuity conduit fittings
(a) Old types   (b) New types

junctions, but in any case the cable space should not exceed 40% of the conduit space.

The wiring capacities stated in Table 3 relate to conduit runs with not more than two 90° standard bends. More arduous conditions with more bends may involve a reduction in conduit capacity or an increase in the number of inspection fittings or draw boxes.

The effect of bunching a number of wires together is to reduce the amount of current each wire can carry, as indicated in Table 4. Thus four 4 mm² (7/0·85) cables require 20 mm conduit (from Table 3), and the permissible current is 19 A (from Table 4).

Conductor of cable (p.v.c.)		Heavy-gauge conduit				
Nominal cross-sectional area	Number and diameter (mm) of wires	16 mm	20 mm	25 mm	32 mm	38 mm
mm²		Maximum number of single-core p.v.c.-insulated cables				
1·0	1/1·13	7	10	21	35	—
1·5	1/1.38	6	9	18	31	—
2·5	1/1·78	5	8	14	24	34
4·0	7/0·85	3	4	9	16	23
6·0	7/1·04	2	3	7	12	17
10·0	7/1·35	—	2	4	7	11
16·0	7/1·70	—	—	3	5	8

Table 3 Wiring capacities of steel conduits

Note: The capacities for other elastomer-insulated cables are about 75% of the figures given above.

These current ratings are for single-core, p.v.c.-insulated cables and sheathed cables run bunched, and enclosed in one conduit, troughing or casing.

Plastic conduits are increasingly being used for installations in housing and commercial premises, though not in factories. Table 1 shows them to compare favourably with other good-quality systems. Plastic conduit is much more flexible than steel conduit; it is more easily bent to shape; and although the heavy-gauge tubing can be screwed, if required, it can simply be cemented into the spouts of the associated accessory fittings and boxes. Some manufacturers

Number and diameter (mm) of wires	Two cables d.c. or single-phase a.c.	Four cables d.c. or single-phase a.c.	Three or four cables three-phase a.c.
1/1·13	11	9	9
1/1·38	13	10	11
1/1·78	18	14	16
7/0·85	24	19	22
7/1·04	31	25	28

Table 4 Current ratings for small house wiring cables

T.Y.E.H.—4

provide special grip outlet fittings and boxes for plastic conduit. Increasing use is now being made of hollow skirtings and door architraves which are designed to provide accommodation for wiring and to enable outlets to be fitted at any point in a room. This is a great convenience for later alterations, additions and extensions to the installation. Copper and aluminium conduits are used where the advantages of one of these metals over steel is of importance, but aluminium is gaining popularity owing to its labour-saving features of lightness and ductility.

Most of what has been said about steel conduits also applies to plastic, copper and aluminium conduits.

Condensation in conduit. In damp atmospheres where changes of temperature occur there is the possibility of moisture condensing inside the conduit. When the conduit is warm the air inside it expands, and on cooling air and moisture are drawn into the conduit. This 'breathing' action can be minimised by making watertight joints at the cable boxes and switches with compound. The necessity for good continuity in metal conduit systems with either screw or grip fittings presupposes that all joints are clean and metal-to-metal. It has been found that red-lead and tow joints give almost as good electrical continuity as dry joints and have the advantage that they are watertight. Metallic paint is also commonly used to make sound joints. Where the conduit is buried and cement is floated on or much wet plaster is used, these methods are important for preventing the ingress of water into the conduits. The moisture tends to rust the inside of the conduit and is likely to be more troublesome on light-gauge than on heavy-gauge conduit, especially if the enamelling is poor. With long vertical runs, an open T-piece is sometimes fitted at the bottom to allow drainage and some ventilation, while horizontal runs are given a slight slope to the lowest point, where a through box or T is fitted, or a drain hole is provided. Rusty patches and moisture also tend to affect the v.i.r. insulation of cables. For this reason, conduit systems erected during building should be given some time to dry out or thoroughly swabbed before the cables are drawn in.

Continuity and earthing. Metal conduit systems must be continuous electrically and mechanically, and must be effectively connected to earth. The wire used for earthing must have a cross-sectional area related to that of the largest cable used and must not be smaller than 6 mm^2 (7/1·04 mm). Earthing leads must be insulated and coloured green. Special earthing conductors, consisting of single or stranded conductors with green p.v.c. sheathing for electrical and mechanical protection, are also used. Non-metal sheathed wiring generally contains an earth-continuity conductor, which must be connected, together with other earthing wires, to an earthing terminal at all terminations. In plastic conduit systems, an earth-continuity conductor should be drawn into the conduits with the circuit wiring. The sizes of separate earth-continuity conductors are related to the circuit cable sizes: 1 mm^2 e.c.c. is used with circuit wiring up to 2·5 mm^2; 2·5 mm^2 e.c.c. with circuit wiring up to 6 mm^2; and 6 mm^2 e.c.c. with circuit wiring up to 16 mm^2.

Earthing clips are employed for steel conduit systems; they must be put on bare metal, so the enamel has to be scraped off the conduit. They consist of tinned brass or copper with a terminal or lug, to which the earthing wire is screwed or soldered. An earthing terminal is provided on metal-cased apparatus. Continuity and earthing are safety precautions, so that in the event of the conduit or other exposed metalwork becoming 'live' the fault current can leak to earth. With a 'dead earth' or short-circuit the fuse will blow, isolating the faulty circuit, while small faults may only show up on an insulation test. With poor joints the contact resistance will be high, and may constitute a fire and shock risk if the fault current is not sufficient to operate the circuit protective device; it is there-

fore specified that the resistance between the end of the earth–
continuity conductor or steel conduit at the main switch
and any other point on the installation shall not exceed 1 Ω.
(See also 'Earthing for safety' in Chapter 4, page 92.)

2 Sheathed wiring

In lead-covered wiring the conductors are insulated with
vulcanised rubber, the cores being of the usual colours, and a
sheath of extruded lead alloy encases the insulation. This was
once a popular method of installation, as the wiring was easily
run under floors and in plaster walls. Care had to be taken
that the plaster did not cause chemical damage to the lead
sheath. A disadvantage was that nails could easily pierce the
lead and short-circuit the wires.

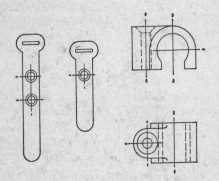

Fig. 41 Plain buckle clips and nylon clip

P.v.c. (polyvinyl chloride compound) sheathed cables
have now entirely displaced lead-sheathed wiring systems in
new installations. The conductors of p.v.c.-sheathed cables
are insulated with red and black p.v.c. to distinguish phase
and neutral wires; and two or three cores, placed side by side,
with or without a bare earthing conductor, and sheathed
overall with p.v.c., are commonly used for house wiring. It
is important to use an ample number of fixing clips to avoid

sagging of the cable and to preserve a neat appearance. Although the p.v.c. is very tough, it is vulnerable to mechanical damage. As mechanical protection, oval conduit or a channel of wood, metal or plastic is used for running down walls. Plastic channelling is shown in Fig. 43. Fixing clips are illustrated in Fig. 41. The buckle clips are fixed by brass nails and bent over the cable, the end being threaded through the eye. Modern cable clips are made of nylon or some other hard plastic and designed to hold the cable when it is pressed into place. A junction box for p.v.c.-sheathed cables is shown in Fig. 42. Harness systems consist of a central junction box with suitable lengths of cable to reach all the outlet points in a house, flat or bungalow and are factory-made ready to install, thus saving much work on site. This form of wiring is economical when made in large quantities for housing estates. A bond wire (or earth wire) is usually run in metal- and p.v.c.-

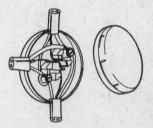

Fig. 42   Junction box for p.v.c.-sheathed cables

Note earthing terminal in centre.

sheathed cables, and consists of a separate, uninsulated conductor. It is important to:

(a) Connect all earthing conductors together, and to an earthing terminal in the junction box, switch box or other outlet accessory where each cable is terminated.

(b) Take the sheathed cable end right into the outlet enclosure so that the cable cores, where unsheathed, are contained within and not exposed outside the outlet enclosure.

There are various patented systems using special fittings which make for ease in erection and safety in operation. Skilled labour is required to make a neat job. Cable should be carefully run off the drum, as kinks are very difficult to get out once they occur and are liable to fracture the conductors.

A system that was used before the introduction of p.v.c. insulation and sheathing was tough rubber-sheathed wiring (t.r.s.), which consisted of vulcanised india-rubber-insulated cores with an additional outer protection of tough vulcanised rubber. Care must be taken to avoid sharp bends and undue

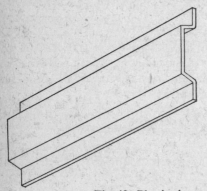

Fig. 43 Plastic channel

pressure on plastic sheathing. It is advisable to keep these types of cable out of direct sunlight to avoid softening.

In multi-core cables the earthing conductor is usually bare within the sheathing, but where it is exposed at terminations it should be insulated with green sleeving. Although insulated lampholders and insulated switches do not need earthing, it is a rule to provide an earthing terminal in all outlet boxes so that any exposed metalwork of future fittings and switches, if changed, can be earthed.

P.v.c. cables are tough and flexible at normal operating temperatures, highly resistant to burning, chemically inert and unaffected by sunlight. As p.v.c. is thermoplastic it should not be used under heat and pressure. The current ratings are published in the I.E.E. Regulations. For conduit work p.v.c. is moulded directly onto the wire and does not need further protective coverings.

3 *Cleat system*

At one time this was often used as a temporary system in temporary buildings for economy and rapid construction, but p.v.c.-sheathed wiring is now commonly used for both temporary and permanent wiring as it is simpler and easier to install, the only difference being in the degree of care necessary to make the installation neat in appearance. With the old cleated system, single cables were used; they had to be visible throughout the run and be supported away from the wall by porcelain cleats, which are illustrated in Fig. 44 and should be fitted out of reach.

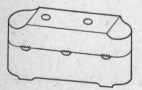
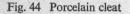
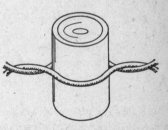

Fig. 44 Porcelain cleat   Fig. 45 Knob insulator

Additional protection, such as porcelain or metal tubes, is required when passing cables through walls or floors. The wiring must be kept clear of gas and water pipes and any other metalwork.

Cleat wiring is seldom seen nowadays. Flexible twin wires

on knob insulators, shown in Fig. 45, are to be found abroad, but this system is not employed in this country as it is not considered safe or robust enough.

4 *Casing and capping*
Many of the earlier installations were run in a wood casing with a moulded capping, as illustrated in Fig. 46. This system

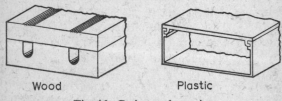

Wood   Plastic

Fig. 46   Casing and capping

was expensive to install and necessitated highly skilled labour to make a good job. The modern form of this system is black or white rectangular-section plastic cable trunking of various sizes with clip-in covers to contain p.v.c.-insulated single wires; it is often used in conjunction with a plastic conduit system. The walls should be plugged and the casing secured by screws. The capping is slipped into position by springing the trunking open to allow the tongued and grooved edges to engage.

5 *Mineral-insulated cables*
This type of wiring consists of single- or multi-core, single-strand conductors surrounded by compressed magnesia insulation with an external metal sheath formed into a solid-drawn cable. The conductors and sheath were originally of copper, but because of the high cost of copper aluminium is now being used extensively instead. An outer p.v.c. sheath is added where the metal sheath would be liable to corrosion. This cable is very heat-resistant, even fire-resistant, to a degree beyond all other cables, and can suffer appreciable

SWITCHGEAR, DISTRIBUTION AND WIRING67

deformation without electrical failure. The ends are liable to absorb moisture, so they must be sealed with special accessories to obtain a watertight installation. Fig. 47 shows a typical cable termination with a screwed compression gland. Current ratings for mineral-insulated cables, not p.v.c.-sheathed, are much higher than for p.v.c.-insulated cables, which means that they can operate at higher temperatures, but voltage drop in long runs is consequently a more important consideration. A mineral-insulated cable must not be allowed to reach a temperature exceeding 250° C (482°F) provided that it is out of reach and not likely to be injurious to its surroundings (150°C for terminations).

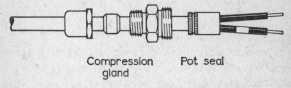

Compression   Pot seal
gland

Fig. 47   Mineral-insulated cable termination

Special systems

A system of wiring similar to that commonly used in America is now available in this country. This is a p.v.c.-insulated, multi-core cable sheathed in flexible corrugated metal tubing which requires special glands to terminate the ends in junction boxes and outlet boxes (Simplex-BICC).

Harness systems have already been mentioned, but they have a special application in housing schemes where the same design and size of 'harness' can be produced in quantities economically in the factory for a large number of houses or flats. Almost any form of sheathed cable can be used for such systems.

One rewireable system consists of small-bore plastic tubes connected by webs beween them, each tube containing a single, loose, p.v.c.-insulated cable.

Concentric wiring or 'sheath return' systems can be used where the neutral conductor of the supply is earthed, and if permitted by the supply undertaking. The outer metal sheath of the cable must be used as the neutral return conductor, the line or phase conductor being the insulated core conductor to which all switches must be connected. The metal sheath must, of course, be electrically continuous and also effectively earthed. For this system the mineral-insulated, metal-sheathed cables are ideal, and these are mainly used where the system is permitted. Some supply undertakings that allow the use of this system require the metal sheath to have an additional p.v.c. sheath.

Looped wiring in a conduit system

The conventional diagram for the parallel connection of three lamps is shown in Fig. 48, from which it appears that joints

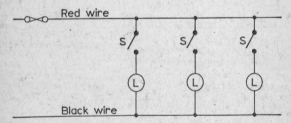

Fig. 48 Conventional diagram for three lamps in parallel, with switches

would have to be made from the red and black feed wires. Such joints would have to be well made and then insulated. This method has long been obsolete. Instead of spending the time and labour on joints, slightly more cable is used, and loops are brought into one side of the switch and lamp, as shown in Fig. 49. Between the other side of the switch and lamp is the switch wire, which should be red. If the lamp is on

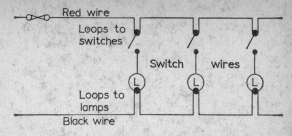

Fig. 49 Looping out to lamps and switches

a fixed bracket or lampholder, the black wire can be looped in and out again, the loop going to one terminal. When pendant fittings are employed, the looping is done at the ceiling-rose instead of at the lampholder, which is connected to the ceiling-rose by a twin flexible cable. This is illustrated in Fig. 50, which shows how three wires enter each switch and the ceiling-rose, but two of them are looped.

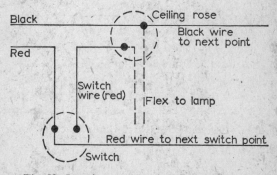

Fig. 50 Looping out to switch and ceiling-rose

When twin sheathed cable is used, it is possible to use a junction box with terminals or cable connectors to enable the lamp to be connected in series with the switch, as shown in Fig. 51. Both red and black wires are fed into a junction box with four entries. Three twin cables are taken out, as indicated in the diagram; one twin cable provides the feed to the next point, the next outlet goes to the ceiling-rose, while the third goes to the switch. The connections within the junction box are made with brass terminals fixed to the base of an insulated

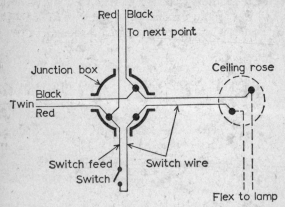

Fig. 51 Twin wiring to junction box, switch and ceiling-rose (earth wire and terminal not shown)

box, or with cable connectors having brass terminals encased in an insulating enclosure if the box is of metal.

Another method in conduit or sheathed cable wiring systems, using a three-plate (terminal) ceiling-rose, is shown in Fig. 52, the circuit connections of which are made with

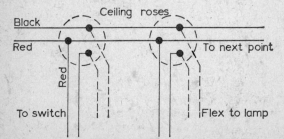

Fig. 52 Three-plate ceiling-rose showing looping out

pairs of wires. This method is not very common, but it is more economical in cable and junction boxes.

In all systems of wiring it is most important to connect a single-pole switch in the line or phase lead to the lamp, and never in the neutral or return lead. Regulations prohibit any break in the neutral conductor, except in a linked switch which isolates both poles of the circuit together.

4 Flexible Cords, Cables and Testing

When apparatus is in a fixed position, permanent cabling or 'hard' wiring of a rigid type can be installed, and for large installations the main current can be carried by bus-bars of copper rod or strip. But in any system provision must be made for connecting lamps, kettles, irons and other portable apparatus to the permanently-wired outlets. For this purpose flexible leads are required, and these flexible leads are one of the most common points of failure, as they are often misused and are liable to be damaged through carelessness. For household use a flexible lead with a pleasing finish is desirable, while workshop flexibles must be more robust and do not require an artistic finish. Flexible leads are colloquially called 'flex', and the conductor is made up of a number of fine wires twisted together and insulated with vulcanised india-rubber, p.v.c. or some other elastomeric insulation. With v.i.r. and elastomeric compounds, the copper wires are tinned to prevent chemical reaction between the compounds and the copper. It is wise to buy good-quality flexible cords. The I.E.E. current ratings for flexible cords are given in Table 5 on page 73.

The style of finish depends on the application and may be braided coloured textile for decorative purposes, or black or white p.v.c for domestic use, or tough rubber for workshop flex.

For larger currents than those given in Table 5, flexible cables from 6 mm^2 to 630 mm^2 are available.

Twin twisted flex consists of two insulated and braided conductors twisted together to form a pair.

Circular flex consists of insulated conductors laid up and

Conductor		Current rating (subject to voltage drop) d.c. or single-phase or three-phase a.c.	Maximum permissible weight supportable by twin flexible cord
Nominal cross-sectional area	Number and diameter (mm) of wires		
mm²		A	kg
0·5	16/·20 or 28/·15	3	2
0·75	24/·20 or 42/·15	6	3
1·0	32/·20	10	5
1·5	30/·25	15	5
2·5	50/·25	20	5
4·0	56/·30	25	5

Table 5 Flexible cords

twisted together with hemp worming, the whole being sheathed circular to suit the type of application.

For domestic use unkinkable flex is recommended. The insulated conductors are surrounded by a rubber compound and heald cords for strengthening purposes, and have an overall covering of textile braid; thus any pull or jerk—as, for example, when the electric iron falls off the table—does not strain the copper conductors or their insulation. The construction of unkinkable domestic flexible is illustrated in Fig. 53. The heald cords can be anchored at the ends if necessary, but most accessories and appliances designed for flexible cord connections have efficient clamps to grip the flexible, as illustrated by the three-pin plugs shown in Figs. 54 and 55 on pages 75 and 76. The regulations require that all portable apparatus, including lampstandards, must be fitted with circular braided or sheathed flexible. Unfortunately, this regulation is not always observed on some, generally foreign, portable apparatus offered for sale to the public.

Three-core flexible

The rule regarding earthing of portable apparatus states that all exposed metal, with the exception of certain minor iso-

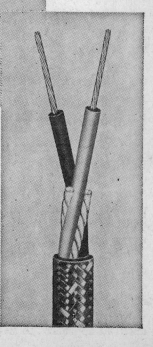

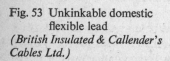

Fig. 53 Unkinkable domestic
        flexible lead
(*British Insulated & Callender's
Cables Ltd.*)

lated parts, must be earthed. For this purpose a three-core flex is used, which consists of three insulated conductors. The live core is coloured red or brown and must go to the switch, if one is fitted on the appliance; the black or blue core is the neutral; and the green or green/yellow striped core goes to the

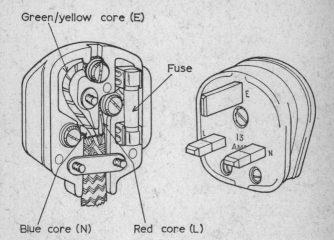

Fig. 54 Three-pin plug (flat pins) and flexible lead
(M. K. Electric Ltd.)

earthing terminals of the appliance and the three-pin plug. (The former are the old colours and the latter are the new standard colours.) The earthing-pin is the larger of the three pins and is generally arranged to fit in the top of the socket outlet. The method of attachment and connections are shown in Figs. 54 and 55.

Prior to 1970 the established colours of cores in flexible cords were red for the 'live' conductor, black for the neutral conductor and green for the earthing conductor, but international standards of colouring have now been adopted which avoid the confusion and danger of connecting imported electrical appliances to flexibles having different-coloured

Green/yellow core (E)

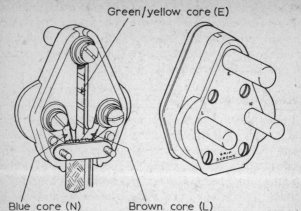

Blue core (N)   Brown core (L)

Fig. 55 Three-pin plug (round pins) and flexible lead
(M. K. Electric Ltd.)

cores. The correct colours are now brown for the 'live'
conductor, blue for the neutral conductor and green/yellow
stripes for the earthing conductor.

Tough rubber- and p.v.c.-sheathed flexible

Sheathed flex is employed for trailing cables, in damp positions and where good mechanical protection is essential.
T.r.s. sheaths are the most flexible but are affected by greases
and oils. P.v.c. sheaths, however, are oil- and grease-resistant,
and are also easily cleaned.

Special flexibles for very arduous duties and use where
subject to abrasion are protected by a wire braid armour over
the sheathing.

Pendant and other flexibles

Parallel twin p.v.c.-insulated flexible may only be used for
fixed lighting fittings or where it will not be subject to abrasion or undue flexing. Twisted twin non-sheathed flexibles

insulated with rubber or p.v.c. may be used in the same way without limitation on flexing and, in addition, for pendants that are entirely open to view. Twisted twin non-sheathed flexible insulated with transparent p.v.c. may only be used for pendants that are entirely open to view.

Heat-resistant flexible

All electrical equipment generates heat when in operation, and it is one of the manufacturer's design problems to dissipate the heat without excessive temperatures developing inside or on the outside of the appliance, whether it is a lighting fitting or a cooker. The flexible lead to a light fitting or a portable heating appliance passes through hot surfaces and is in hot surroundings where it enters the casing or enclosure; heat is injurious to rubber and thermoplastic insulation and sheathing. All lighting pendant flexibles are subject to excessive temperatures where they emerge from the lampholders. Heat-resistant flexibles should therefore be used for all such applications. The common heat-resistant insulating materials for flexibles—in ascending order of heat-resisting quality—are: butyl or ethylene propylene rubber and heat-resisting p.v.c., varnished p.t.p. fabric, silicone rubber and varnished glass fibre. The last-named should be used for the lampholders of tungsten lamps and lighting fittings, and the former types of insulation for block storage heaters, fires and immersion heaters.

Sizes of conductors

The current ratings and consequent sizes of cables and wiring conductors depend on whether the cables are to be run singly or bunched together, and whether they are supported in air or enclosed in a conduit or trough, as already explained.

Other considerations that affect the size to be used are:

(a) The voltage drop, due to the resistance of the conductor and the current it has to carry.

(b) The heat-dissipating qualities of the type of cable insulation to be used.

(c) The minimum size of wire to be permitted.

Voltage drop in cables

The effect of incorrect voltage has been dealt with in an earlier section (see p. 39), but here we are concerned with the effect of the cable sizes chosen on the voltage available at the terminals of the lamp or other apparatus.

The rule for the maximum allowable voltage drop is $2\frac{1}{2}\%$ of the supply voltage.

With the standard voltage of 240 this is:

$$\frac{2 \cdot 5}{100} \times 240 = 6 \text{ V}.$$

On a 110 V circuit it is 2·75 V, and as the current per watt is greater much larger sizes of cable must be used to keep within the permissible voltage drop.

Of this figure, about 2% drop should be spread over the main cables and those forming the sub-circuits from the distribution boards in about equal proportions for economy in wiring. The remaining $\frac{1}{2}\%$ is left for the voltage drop at contacts of distribution boards, switches, etc. In long runs of mains it is desirable to keep the cable sizes larger than those required by the allowable voltage drop. This allows for any future extensions and growth of load, and also minimises the effect of lower supply voltage, which may occur at times of peak load on the supply network. With large loads such as 3 kW electric fires or water heaters at the end of a long run, the relatively large current taken by such apparatus affects the voltage of all the sub-circuits, so that some liberality in cable sizes is justified. Table 6 on page 79 gives some of the sizes of various cables used in house wiring. For more complete lists, reference should be made to the Institution of Electrical Engineers' *Regulations for the Electrical Equip-*

ment of Buildings, 14th Edition 1966, with later amendments and 1969 Metric Supplement, or to books on electrical installation work.

The rated currents given apply at an ambient air temperature of 30°C (86°F). Above this temperature, or with larger groups of conductors, the current rating is reduced as specified in the Regulations. Paper- or cambric-insulated, lead- or aluminium-sheathed cables have better heat-dissipating qualities and can carry much greater currents than those given in Table 6.

Conductor		Two cables d.c. or single-phase a.c.		Three or four cables three-phase a.c.	
Nominal C.S.A.	Number and diameter (mm) of wires	Current rating	Voltage drop per ampère per metre run	Current rating	Voltage drop per ampère per meter run
mm²		A	mV	A	mV
1	1/1·13	11	40	9	35
1·5	1/1·38	13	27	11	23
2·5	1/1·78	18	16	16	14
4	7/0·85	24	10	22	8·8
6	7/1·04	31	6·8	28	5·9
10	7/1·35	42	4·0	39	3·5
16	7/1·70	56	2·6	53	2·2
25	7/2·14	73	1·6	71	1·4
35	91/1·53	90	1·2	88	1·0

Table 6 Current ratings for single-core p.v.c.-insulated cables, non-armoured, with or without sheath (copper conductors), run bunched, and enclosed in conduit, troughing or trunking, with ambient temperature of 30°C and where coarse excess-current protection is provided

A further modification in the current-carrying capacities of cables depends on the class of excess-current circuit protection provided for the cables. Table 6 gives values for coarse-current protection, which is provided by most semi-enclosed (rewireable) fuses. With close excess-current protection, such as that provided by high rupturing capacity (h.r.c.) cartridge fuses and miniature circuit-breakers, the current-carrying capacities of the cables can be raised by 33 % because the close excess-current protection will operate quickly (within 4 hours with 50 % excess current), whereas coarse protection will take longer to operate, allow a dangerous fault current to persist for too long and cause damage by overheating the cable.

Choice of cable size

It will be evident that the maximum permissible current that any cable can carry will depend on a number of factors that must be applied to the basic figures in tables. These factors are given in appropriate tables for various types of cable in the I.E.E. Regulations, and they cover type of protection, grouping, conditions of installation and ambient temperature. The current required can be estimated and the most suitable size of cable selected. For small house installations the smallest appropriate cables are usually adequate for mains, lighting and socket outlet sub-circuits without regard for voltage drop or other factors involved, but where there are long runs the voltage drop should be checked to see that it is satisfactory. The minimum size of conductor for sub-circuits is 1 mm², which is 1/1·13 mm diameter strand rated at 11 A, and with flexible cords the minimum size is 0·5 mm², which consists of 16/0·2 mm diameter strands and is rated at 3 A.

Before metrication in this country, the two available single-strand cables had been found so stiff and difficult to draw into conduit, and a fracture would be so serious, that the smallest three-strand and more flexible cable was almost always used instead, although slightly larger in size. Metrication, however, has resulted in the introduction of three sizes of single-strand cable which are all larger than the previous two and even more difficult to handle. Although this is an advantage for sheathed wiring, it remains to be seen whether pressure from the contractors will force the manufacturers to introduce smaller and more flexible stranded cables. The resistance of small cables is given in the table below for the purpose of voltage drop calculations.

No. and diam. of wires	1/1·13	1/1·38	1/1·78	7/0·85	7/1·04
Ohms/km at 20°C	17·1	11·6	6·92	4·7	2·97

Table 7

Voltage drop calculations

From Ohm's Law we know that the voltage drop $= I \times R$, where I is the current in ampères and R is the resistance in ohms. From the load in watts (P), the current (I) can obtained by dividing by the voltage (V), i.e. $I = \dfrac{P}{V}$. The resistance per kilometer of various cables is given in wire tables, so the 'voltage drop' can be calculated. All the 'drops' of conductors in series added together, from the service up to the apparatus concerned, should not exceed $2\frac{1}{2}\%$ of the declared voltage.

Example 10. The load in a house is 7 kW and the supply voltage is 240. If the distance from the incoming supply to the distribution fuseboard is 10 m, what will be the voltage drop and the size of the two single-core main cables?

$$\text{Current } I = \frac{P}{V} = \frac{7 \times 1000}{240} = 29 \cdot 2 \text{ A.}$$

Referring to Table 6, two single-core cables of 7/1·04 mm in conduit or casing will carry 31 A. The total length of the cable, flow and return is $2 \times 10 = 20$ m, and its resistance per kilometre is 2·97 Ω. Thus the total resistance (R) is $\frac{20}{1000} \times 2·97 = 0·059$ Ω. Hence the volt drop is $Vd = IR = 29·2 \times 0·059 = 1·72$ V.

This is well within $2\frac{1}{2}\%$ of the supply voltage (6 V), but there will also be some voltage drop in the final sub-circuit cables.

With a number of lamps and other apparatus, the total load is distributed over the length of sub-circuit cable from the distribution board. The farther away a point is from the distribution board, the less is the current but the greater is the cable resistance. Thus the total load can be considered as concentrated at some point along the sub-circuit cable. This 'load-centre' may be the geometrical midpoint of the group, or it can be found in the same way as a centre of gravity (or centroid) in applied mechanics. With a well-distributed load it can be assumed to be halfway along the sub-circuit route. To find the percentage voltage drop measure the distance from the distribution board to the load centre, add up the maximum load on the branch circuit in watts and work out:

$$\text{Percentage voltage drop} = \frac{PRL}{10V^2},$$

where P = load on the circuit in watts
R = resistance per kilometre of the conductor used,
L = length of lead and return, i.e. twice the route length in metres,
V = declared supply voltage.

Example 11. The lamp load on a circuit is 650 W, and the route length is 30 m. If 1/1·13 cable, having a resistance of 17·1 Ω per kilometre, is used, what is the percentage voltage drop on a 240 V supply?

Using the formula above:

$$\text{Percentage voltage drop} = \frac{650 \times 17\cdot1 \times (2 \times 30)}{10 \times (240 \times 240)} = 1\cdot16\%$$

so the actual voltage drop is $\frac{1\cdot16}{100} \times 240 = 2\cdot78$ V. This, added to the voltage drop in the main cables (1·72 V), makes a total of 4·5 V, which is well within the permitted $2\frac{1}{2}\%$, so the cable size is satisfactory.

There will be a slight difference between the values of voltage drop calculated from the conductor resistances and those worked out from the volts drop per ampère per metre run given in Table 6. This is because the former are standard values for cable conductors at 20°C, whereas the latter are values for cables in an ambient temperature of 30°C carrying the rated current. Complicated temperature correction factors are involved to account for the differences. However, for all practical purposes it is simpler and satisfactory to use the voltage drop figures in the cable current tables, on which Table 8 is based for a group of two cables. This shows the maximum length of run in metres for 1 % voltage drop on a 240 V system with various load currents for five small cables.

Load in W		6000	5000	4000	3000	2000	1500	1250	1000	750	600	500
Current in A		25	20·8	16·7	12·5	8·33	6·25	5·2	4·16	3·12	2·5	2·08
Conductor												
Size	Rated current (A)											
1/1·13	11	Maximum permissible current exceeded				7·2	9·6	11·6	14·4	18	24	29
1/1·38	13				7·1	10·7	14·2	17	21	28	36	43
1/1·78	18			9	12	18	24	29	36	48	60	72
7/·085	24		11·5	14·4	19	29	38	46	58	77	96	116
7/·104	31	14·1	17	21	28	42	56	68	84	112	140	169

Table 8   Maximum length of run, in metres, for 1 % voltage drop on a 240 V supply

These figures are approximate as the voltage drop basis in the cable current-carrying capacity tables is only correct for the full rated current of the cable; corrections for temperature at different loads, which affects the conductor resistance, would be necessary for accurate figures.

Effect of a heavy load on the end of a sub-circuit

Fig. 56 (a) shows a pair of main cables, each with a resistance 0·05 Ω and a sub-circuit with one 100 W lamp halfway along it and another 100 W lamp at the end. The resistance of each sub-circuit wire is 1 Ω. With the first 100 W lamp switched onto the 110 V supply, a current of $\frac{100}{110}$ = 0·91 A flows along the main cables and half the length of the sub-circuit. The total resistance is 0·05+0·5+0·5+0·05 = 1·1 Ω, so the *IR* drop will be 0·91×1·1 = 1 V. Thus the voltage at the first lamp will be 109 V. When the second lamp, at the end, is switched on, the total current will be 1·82 A from the mains. The voltage drop is 1·82 × 1·1 = 2 V at the first lamp plus 0·91 × 1 = 0·91 V at the second lamp. Thus the voltage across the second lamp is 107·09, say 107 V. This illustrates how the lights may be less brilliant as one proceeds along a long sub-circuit. Now, suppose the occupier thinks he would like to have a 1 kW fire in the same room as the last lamp and connects the fire to the end of the sub-circuit (see Fig. 56 (b)).

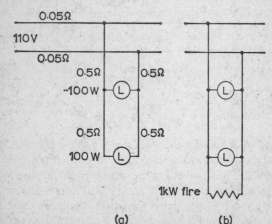

(a)   (b)

Fig. 56   Illustrating voltage drop and heavy load on long sub-circuit (circuit fuses and switches not shown)

This fire will take $\frac{1000}{110} = 9\cdot1$ A, which is added to the existing lighting load and was never contemplated when the wiring was put in. Now let us work out the voltage across each lamp.

The total current from the mains is $9\cdot1+1\cdot82 = 10\cdot92$ A. Drop to first lamp $= 10\cdot92\times1\cdot1 = 12\cdot01$, say 12 V, so the first lamp has only 98 V across it, or 89% of its proper voltage. In the remaining half of the sub-circuit the current is $9\cdot1+0\cdot91$, say 10 A, and the further drop in voltage is $10\times1 = 10$ V; so the second lamp has only $98-10 = 88$ V across it, which is 80% of its proper voltage, and thus it will burn dimly. These abnormally high resistance figures have been chosen deliberately to bring out the effect of a heavy load on

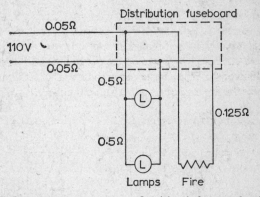

Fig. 57 Correct way to connect a fire (circuit fuses and switches not shown)

the end of a sub-circuit, which is not good practice. The correct way is to run a separate circuit for the electric fire. This is shown in Fig. 57, in which the resistance of the wires to the fire are each taken as $0\cdot125$ Ω. We will now calculate the currents and voltages across the lamps. The main cable still carries $10\cdot92$ A and the drop is $1\cdot09$ V. The drop in the wires to the fire is $9\cdot1$ A $\times0\cdot25 = 2\cdot27$ V, so the fire is on $108\cdot9-2\cdot27 = 106\cdot6$ V, which is not too bad. The voltage drop to the first lamp is 1·82 V, giving $110 - 1·09 - 1·82 = 107$ V, while the second lamp is on $107 - 0·91 = 106$ V (approx.). This compares favourably with the figures of 98 V and 88 V respectively in the former case.

Heat and power circuits

From the above example it can be seen that power and lighting circuits should be run separately. In radial final sub-circuits it is generally sufficient to choose the cables for the circuits that supply electric fires and other apparatus on a current-carrying basis, as the current taken is separate from the lighting circuits and the voltage drop is less important. A 2 kW fire on 110 V takes 18·2 A, so that 7/·085 could be used. The same size of fire for 240 V takes 8·3 A, so 1/1·13 is large enough. But there is a possibility of a larger fire being fitted, say 3 kW, at some later date; and if a 15 A socket-outlet is provided, then 1/1·78 to carry 18 A should be used.

For apparatus taking up to 1 kW on 200–250 V, a 5 A plug and socket can be used, and with nothing else connected to this circuit 1/1·13 has enough current-carrying capacity. For larger currents between 5 A and 15 A, there is only the 13 A or 15 A plug size, e.g. 2 kW at 200 V takes 10 A; so a 13 A or 15 A outlet is employed and the rule is to run cable to match the outlet, i.e. 1/1·38 for the 13 A socket or 1/1·78 for the 15 A socket is used. But radial final sub-circuits for 13 A socket outlets may supply up to six points (with the exception of fixed water-heating points) in one room of less than 30 m² area, except a kitchen, or two points in other rooms, with 2·5 mm² (1/1·78) p.v.c. cables protected by a 20 A fuse; or up to six points in other rooms with 4 mm² (7/0·85) p.v.c. cables with a 40 A fuse. The corresponding alternative mineral-insulated cables are 1·5 mm² and 2·5 mm² respectively.

To determine the rating of a final sub-circuit for stationary cooking appliances the total current is assessed by taking the first 10 A of the total rated current of the appliances, plus 30 %

of the remainder, plus 5 A if the cooker control unit incorporates a socket outlet.

The current rating of all other circuits is taken as the sum of all socket outlet ratings, and stationary appliances are rated at their normal current.

The ring circuit

In domestic installations it has become common practice to provide an ample number of 13 A socket outlets for all purposes throughout the house, with several in each room, and for these different rules apply. This great convenience means that only a small proportion of the number will supply heavy-current appliances at any one time, so an economical

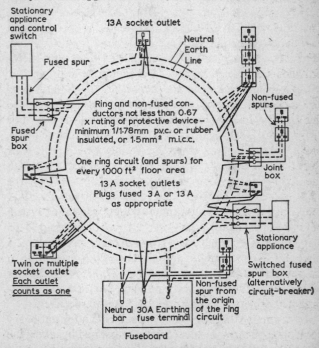

Fig. 58. Ring circuit design *(A.S.E.E.)*

wiring arrangement has been devised called the ring circuit. The wiring forms a ring with the ends connected to a 30 A fuse, the wiring being looped into each point. Each ring may serve an unlimited number of 13 A socket outlets within a maximum area of 100 m². 2·5 mm² (1/1·78 mm) cables are used for the ring circuit. The circuit fuse limits the total loading to 7·2 kW (at 240 V) and the small cartridge fuses of 3 A or 13 A capacity in the plugs protect each standard-lamp or other appliance connected to the ring circuit independently. A limited number of spurs of the same sized cable may be taken from the ring to not more than two 13 A sockets on each spur. The ring circuit is shown in Fig. 58.

Lighting circuits

The current rating of lighting final sub-circuits is based on the connected load, but not less than 100 W per lampholder, and all the other small points such as electric clocks, shaver units, bell transformers, etc., with ratings less than 15 VA can be neglected. In domestic premises either 1/1·13 mm or 1/1·38 mm is usually large enough, and the latter will, of course, have less voltage drop. Any 5 A sockets should be taken at their full value, and 2 A sockets can be taken at $\frac{1}{2}$ A since, in general, these are only used for lighting purposes.

Diversity of installation loads

In medium-sized houses it is not usual for more than half the lighting to be on at the same time, though two-thirds is taken for calculating the maximum demand of the premises. This should not lead to a reduction in cable size, as the wiring is already small enough for mechanical reasons.

For heating and power appliances, the full load up to 10 A is taken, plus 50% of any load in excess of this value. The cooker load is taken as already described, and water heaters (thermostatically controlled), floor warming and thermal storage space heating installations are taken at their full rated currents. The loading of 13 A general-purpose socket outlet circuits is taken as the full rating of the largest fuse or circuit-breaker rating of the individual circuits (which is 30 A for a ring circuit), plus 40% of the sum of the fuse or c.b. ratings of the other circuits. For other socket outlets and stationary appliances, a similar method applies, i.e. the rating of the largest point, plus 40% of the sum of all remaining point ratings. All sections of the installation, so calculated, are added together to obtain the maximum load that will be demanded from the supply undertaking and carried by the main cables and switchgear. In larger buildings and institutions, different proportions are employed. These values should not be applied in assessing the loading of distribution boards as they only represent the diversity of demand to be expected in a complete installation.

Installation testing

In the complete installation all the wiring is tested for continuity and insulation resistance. It is a rule that the earth continuity shall not exceed 1 Ω resistance generally, but 0·5 Ω is the maximum for low-current testing of steel conduits used as the earth-continuity conductor. The insulation resistance of the system depends but slightly on the grade of cable used, provided that there are no breaks in the insulation covering the wires. Most of the leakage current occurs at cable ends, switches, distribution boards and points where the connections are made. The total insulation resistance *decreases* with increase of length of cable runs and with more outlets. The rule is that the insulation resistance of the wiring shall not be less than 1 MΩ between line and neutral conductors, and between conductors and earth. The insulation to earth of fixed apparatus that it is practicable to disconnect for the tests may be as low as 0·5 Ω. To test the wiring an ohmmeter is used. A trade name for such an instrument made by Messrs Evershed & Vignoles is the 'Megger', but this name is often applied in practice to all such instruments. The instrument consists of a hand-driven generator which supplies a constant voltage to the installation under test. A voltage commonly employed is 500, i.e. twice the working voltage of the installation, though higher values are obtainable. There are two terminals on the 'Megger', one labelled earth and the other line. When the handle is driven, a constant speed is ensured by means of a slipping clutch. The pointer on the instrument moves over a scale of ohms, which varies from infinity to zero. For the larger values megohms, or millions of ohms,

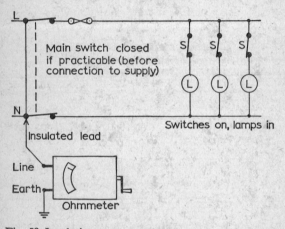

Fig. 59 Insulation test to earth before supply is connected

are marked on the scale. More modern types of insulation and continuity testing instruments use electronic internal circuitry and small dry batteries for their current supply.

With a perfect insulator, the pointer goes to infinity, while with a perfect conductor, i.e. perfect continuity, the pointer goes to zero. Thus this instrument can be used for testing insulation resistance or continuity.

Two tests of insulation are performed, one 'to earth' and the other 'between conductors'. Fig. 59 illustrates the insulation test to earth, which is made with all lamps and fuses in, switches on and all poles or phase conductors connected together (not applicable to earthed concentric wiring). The line terminal of the 'Megger' is connected to one conductor, while the earth terminal is connected to a good earth, such as a metal water pipe that enters the ground (*not* a gas pipe) or the supply undertaking's earthing terminal. The second test checks the insulation resistance between the two conductors, during which the lamps are removed and all apparatus is disconnected, as shown in Fig. 60.

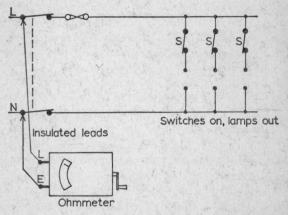

Fig. 60 Insulation test between poles before supply is connected

These tests are usually carried out by the supply undertaking before the installation is allowed to be connected to the service cable. They are a check on the safety of the wiring; but such a test, even if quite satisfactory, is not in itself a guarantee of the safety of the whole installation, for which further tests are necessary as described in the next section. It is important to note that no tests other than earth-loop impedance tests should be made on an installation unless the supply is disconnected and all wiring is dead.

It is recommended that the installation be periodically

inspected and tested at intervals of not more than five years, and an inspection and test certificate should be issued as described in the I.E.E. Regulations.

Earthing for safety

The efficient earthing of the metallic casing of electrical appliances is a recognised safety precaution.

Conduit, iron switch boxes, portable apparatus, cookers, water heaters, washing machines, motors, etc., should all have a proper earth connection. This is often provided by a three-pin plug and socket or a separate earth wire, which should be protected if liable to mechanical damage and must be run back to a good earthing point.

In the event of any insulation becoming defective so that a current-carrying conductor comes in contact with the casing, the earth connection provides a low resistance path to earth and the large fault current will open the circuit by operating the fuses or circuit-breaker, thereby preventing an electric shock to anyone touching the defective apparatus. About 50 V d.c. or 30 V a.c. is the maximum pressure that a human being can safely withstand, but about half this value is the safe limit for animals. Alternating current is more dangerous than direct current, but the effects depend on the physical condition of and the positions of the contact points on the person receiving an electric shock.

The resistance of the earth-continuity conductor in any final sub-circuit must not exceed 1 Ω. This is tested by connecting the earth-continuity conductor in series with one of the circuit conductors, testing the loop and subtracting the known resistance of the circuit conductor or such other cable used for this purpose, as shown in Fig. 61.

Fatalities have occurred to persons using unearthed or faulty apparatus in bathrooms, and special precautions should be taken to see that any exposed electrical metalwork and/or any other exposed metalwork liable to be in contact

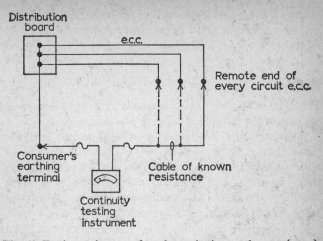

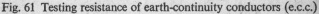

Fig. 61 Testing resistance of earth-continuity conductors (e.c.c.)

with electrical metalwork is properly earthed. The bath and water pipes may have to be bonded to earth in these circumstances. Socket outlets in bathrooms are unsafe and are not permitted, with the exception of special shaver supply units, the sockets of which are entirely isolated from the mains supply by means of a double-wound transformer. Lampholders in the bathroom must be out of reach of a person using the bath and must be of the all-insulated shrouded type. Any switches inside a bathroom must be out of reach of a person using the bath. Alternatively, these can be fixed outside the door, or ceiling switches with insulating cords can be used. These rules apply to any rooms containing a fixed bath or a shower. Fig. 62 shows the various ways earth continuity is achieved in an insulation.

The following example will show an earthing connection that is worse than useless.

Example 12. The end of a 1 kW electric fire element adjacent to the 'live' conductor develops a fault to the metal casing of the fire. The resistance of the fault is 5 Ω and that of the earthing circuit 15 Ω. The supply is 240 V and the fuse fitted will blow at 20 A. Is this apparatus safe and properly protected?

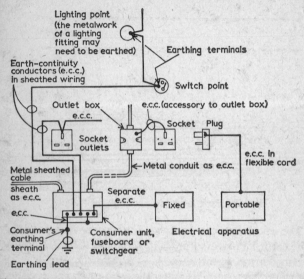

Fig. 62 Methods of achieving earth continuity with earth-continuity conductors (e.c.c.)

(A.S.E.E.)

The circuit diagram is given in Fig. 63, where the neutral conductor is shown earthed at the sub-station and the earth connection or electrode is denoted by E. The resistance from L to E via the fault X is 20 Ω, so the fault current is $\frac{240}{20} = 12$ A. This circuit is in parallel with the fire element, which takes a current of $\frac{1000}{240} = 4\cdot2$ A approximately. Therefore, the total current is 16·2 A, which will not blow the fuse or overload the cable, assuming this is 1/1·78 mm to carry 18 A. But there is 240 V from L to E, as the resistance between E and the sub-station earth is zero, and three-quarters of this potential difference (in proportion to the resistance) is between the points X and E, so that any person who was 'earthed', say by standing on a damp concrete floor, and who touched the frame would be liable to an electric shock at 180

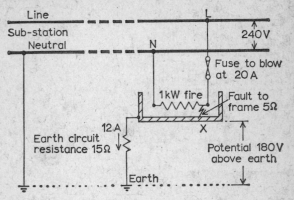

Fig. 63 Earthing circuit of high resistance

V. The calculation can be done another way: the potential *above* earth at point X is the fault current, 12 A, multiplied by the earthing circuit resistance of 15 Ω, namely 180 V.

It will be seen that this apparatus is most unsafe and that the fuse gives no protection against such a fault. Suppose the fault develops into a direct earth of zero resistance and the fault current then increases to $\frac{240}{15} = 16$ A, so that the total current through the fuse is now $16+4\cdot2 = 20\cdot2$ A. The fuse will heat up, and may not blow at once and clear the circuit; but the leakage current will raise the potential of the frame to 240 V—the mains voltage—which is obvious from Fig. 63. Thus the resistance of the earth connection must be well below 15 Ω for the fuse to blow.

Whether the earth-continuity conductor is a separate copper conductor, metal sheathing or conduit, its resistance must be low enough for the fault current to operate the protective device, hence the regulation to limit this resistance.

Earth-loop impedance

Although for the purpose of Example 12 the ground resistance
to the electricity sub-station earth was taken as negligible,
it is not always so. In large towns it is generally very low, but
in some country districts it may be appreciable and may
seriously affect the usefulness of protective devices. Therefore,
it is important to ascertain the earth-loop impedance. Impedance is the same as resistance in d.c. circuits, but with a.c.
circuits it also takes into account the voltage reaction that
may be present. The earth–loop impedance takes into account
the earth resistance itself, as well as the impedance of the
conductors in the sub-station, mains and house wiring,
including the earth-continuity conductors in the earth-fault
circuit.

The earth-loop impedance is tested by passing a current
through this circuit by means of special testing equipment,
which can be plugged into a live socket outlet, and the reading
in ohms must be within suitable limits for the protective
devices installed. The I.E.E. Regulations state that, for semi-
enclosed (rewireable) or small cartridge fuses, the fault
current must not be less than 3 times the current rating of any
fuse for which the ratio $\dfrac{\text{fusing current}}{\text{normal rated current}}$, called the
fusing factor, is greater than 1·5 (coarse excess-current
protection). For cartridge fuses with a fusing factor less than
1·5 (close excess-current protection) the fault current must be
not less than 2·4 times the fuse rating; and for circuit-breakers
the fault current must be not less than 1·5 times the tripping
current of the circuit-breaker.

From these figures the maximum earth-loop impedance
can be worked out for various fuse ratings, as shown in Table
9.

The result of these limitations is to ensure that, in the event
of an earth fault, the earth-leakage current will be high
enough to blow the fuse promptly and so avoid a dangerous
shock voltage existing on exposed earthed metalwork. Thus,

Current rating of fuse	Impedance		
	Fuses		Circuit-breakers
	f.f. > 1·5	f.f. < 1·5	
A	Ω	Ω	Ω
5	16	20	32
10	8	10	16
15	5·3	6·6	10·6
20	4	5	8
30	2·7	3·4	5·4
45	1·8	2·25	3·6
60	1·35	1·7	2·7
100	0·8	1·0	1·6

Table 9 Maximum earth-loop impedances for excess-current protective devices (230–250 V)

with an earth-continuity conductor of 1 Ω resistance or less, there is ample margin for the rest of the earth-fault circuit in the majority of cases. However, where the earth-loop impedance is greater than the permitted figures and cannot be reduced sufficiently by increasing the size of the earth-continuity conductor or by using sufficiently sensitive excess-current devices, then protective gear of greater sensitivity is necessary. These are the current-operated and voltage-operated earth-leakage circuit-breakers, which do not provide excess-current protection and must be used in addition to the normal excess-current protection devices, such as the fuse or excess-current circuit-breaker. They effectively prevent a shock voltage on exposed metalwork from exceeding 40 V, which is considered safe.

The earth-leakage circuit-breaker
An earth-leakage trip may be combined with an overload device, with which it is mechanically interlocked, and this combination can take the place of the main switch and fuse at the minimum of additional cost for the extra protection. But overload (excess-current) and earth-leakage protection

devices are very often separate units so that they can be isolated more easily by the main switch for maintenance purposes.

An automatic earth-leakage circuit-breaker consists of a lightly set circuit-breaker operated by trip coils, which carry either the main load current or the earth-leakage current.

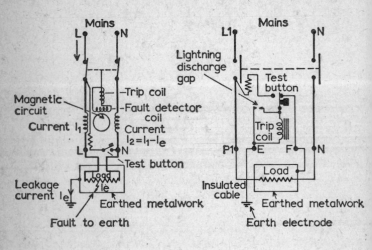

(a)   (b)

Fig. 64 Earth-leakage circuit-breaker connections

(a) Current balance type
(b) Voltage-operated type

The first is called a current-balance e.l.c.b. because two coils are connected in the line and neutral conductors and are linked with a third operating coil in a magnetic circuit, so that earth leakage causes out-of-balance currents to flow through the coils, thus energising the third operating coil of the circuit-breaker; with no leakage, the currents are equal and the third coil remains inoperative. The second type is called a voltage-operated e.l.c.b., in which the operating coil carries any leakage current in the earth-continuity conductor before it goes to earth. The operating coils cause the main contacts to open when energised. The connection diagrams are shown in Fig. 64, and a typical moulded-case earth-leakage circuit-breaker is shown in Fig. 65.

The current-balance e.l.c.b. is commonly used where good earthing is difficult and the earth-loop impedance is too high for any form of excess-current device to afford adequate

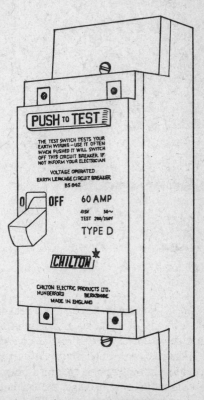

Fig. 65 Typical earth-leakage circuit-breaker
(Chiltern Electric Products Ltd.)

T.Y.E.H.—6

protection. It can operate with leakage currents as low as 500 mA, which permits an earth-loop impedance of up to 80 Ω before the shock voltage can reach 40 V. The voltage-operated e.l.c.b. is the most sensitive type and operates with leakage currents down to about 30 mA. It can be used where the earth-loop impedance is theoretically as high as 1330 Ω. Both types are provided with test resistances and push buttons; they are usually made for load currents up to 100 A, which is ample for the load of most domestic installations.

5 Installation Accessories and Switching

In the last chapter we considered the mains and sub-circuits; now we have to consider the various fittings that may be employed at the ends of the sub-circuits. Good-quality accessories should be used, as the few pence saved on inferior fittings do not compensate for subsequent failures due to the flimsy construction of cheap components.

In the house the majority of the fittings are for lighting, which may be from pendant fittings or wall brackets. Decorative tubular lighting is employed with some interior design schemes, but, as the considerations are purely aesthetic, it is not possible to give any guidance on these applications. In laying out the installation for a new house, it is very desirable to settle the position of the various lighting points before plastering and decoration so that alterations, entailing cutting out and making good, are obviated. With pre-fabricated construction, definite channels are provided for all services, which is an example of production planning, arrangements being made for looping-out to ceiling points for lighting and to socket outlets for power.

Ceiling-roses and plates

The old-fashioned base for a ceiling outlet was a woodblock, which was recessed underneath to allow for the slack ends of the wires. Care had to be taken that this circular ceiling block was screwed securely to a joist or a supporting length of wood secured between joists. The ceiling block was never screwed onto the laths, which tended to split and could not support the weight of a heavy fixture. A porcelain ceiling-rose was screwed onto the block. Modern alternative materials are bakelite and other moulded plastics, and with the development of moulded products such accessories have become common and some types do without the ceiling block. Two or three connectors have grub screws in the sides which grip the looped-in wires, and extension tags carry a round-headed screw and washer. The ends of the flexible leads to the lamp go underneath the washers and the screws are then firmly tightened up. To prevent the pull coming on these terminal screws the separate ends of the flex must be threaded through the porcelain (or bakelite) bridge-piece in between the terminals. These accessories with their small screws and washers, entailing overhead work, do not facilitate quick installation, so more convenient ceiling-roses have been produced in which the flexible connections are made up before the ceiling-rose is assembled on site. A modern, white, moulded ceiling-rose is shown in Fig. 66.

With conduit work ceiling-roses or metal ceiling plates are used that can be mounted directly onto the lugs of a conduit box without using a base block. A ceiling plate with hooks for the supporting chains or tube suspension is employed with some heavy fittings. Bracket fittings should have strong back-plates, and, when the supporting tubes permit, the circuit cables should be brought right up to the lamp-holder. A sufficient length of wire should be left out of the wall for this purpose.

Socket outlets and plugs

The British Standards Institution has laid down standard sizes for sockets and plugs (British Standard Specifications 546 for round pin and 1363 for flat pin types), and non-standard sizes should not be fitted. Nothing is more annoying than to find a variety of plug sizes in a domestic installation, and therefore all sockets for the same purpose in an installation should be of the same size and type; they should also be shuttered. The plugs must go well home into the socket so that no live metal is exposed. Open, live, two-pin sockets are a source of danger, especially to children. Three-pin types usually have shutters actuated by the earthpin. All socket outlets in new domestic installations should be of this

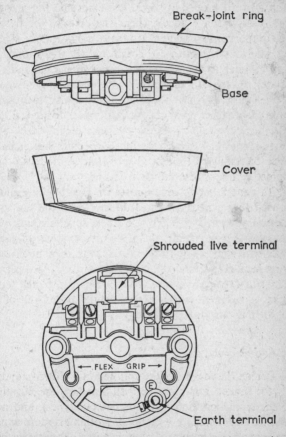

Fig. 66 White moulded ceiling-rose with earthing terminal
(*General Electric Co. Ltd.*)

type, for which it is now standard practice to fit the flat-pin-type fused plugs and sockets to B.S. 1363. One type of three-pin socket has an interlocking arrangement which holds the plug in, unless the switch is in the 'off' position, so that the plug cannot be withdrawn when the socket contacts are live. These however, are not normally fitted in domestic a.c. installations, usually being employed in industry or for d.c. installations.

B.S. 546 standard plug and socket sizes are 2 A, 5 A, 15 A and 30 A, and the wiring to them must be for the rated current to the outlet, no matter what current is taken by the connected apparatus, except that at least 0·5 ampère may be assumed for a 2 A socket. B.S. 1363 fused plugs and shuttered socket outlets are rated at 13 A, with fuses rated at 3 A (red) and 13 A (brown) to B.S. 1362; but 2 A, 5 A and 10 A cartridge fuses are also available, coloured black.

It is common practice for manufacturers to supply 13 A fuses fitted into 13 A plugs, but it is important to replace these with smaller fuses where the plugs are connected to appliances of less than 1500 W rating.

The relative sizes of B.S. 546 round-pin plugs and sockets are shown in Fig. 67. There are several makes of non-standard sockets and plugs on the market—notably those made by Dorman and Smith Ltd. and Wylex Ltd.—which have different arrangements and sizes of pins, but these are not common in modern domestic installations. Wall sockets should preferably be positioned not less than 150 mm from the floor or above working surfaces.

The 13 A socket is particularly intended for the ring circuit described in Chapter 4 and is most widely adopted for house installations. It is made in all forms: to fit onto round and square conduit boxes for surface or flush mounting, with or without a control switch or pilot lamp, and in multiple groups. For d.c. supplies each socket outlet must have a separate control switch because it is not safe to break the circuit by pulling the plug out, due to arcing at the contacts. A switch is not essential for a.c. supplies, although it is very convenient as it saves pulling out a plug to disconnect an appliance. The wiring terminals are made to take 2·5 mm² and 4 mm² looped cables. The twin or triple gang

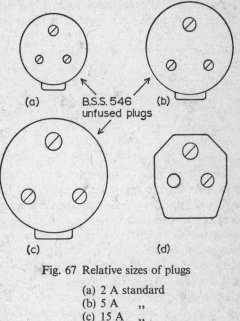

(a)   B.S.S. 546   (b)
             unfused plugs

(c)   (d)

Fig. 67 Relative sizes of plugs

    (a) 2 A standard
    (b) 5 A   ,,
    (c) 15 A   ,,
    (d) Domestic fused plug
               (*Dorman & Smith Ltd.*)

unit is increasingly useful, since it does not involve separate wiring for each socket and, with so many portable appliaances being used in the home, avoids the use of the multiple adaptor. A well-designed installation should be provided with an adequate number of socket outlets so that such adaptors are not necessary, but many installations have such a limited number of socket outlets that the use of multi-outlet adaptors is often inevitable. Fig. 68 shows a typical flush-type 13 A switched socket outlet with neon indicator, and Fig. 69 shows surface-type twin 13 A socket outlets.

For portable lights and small-current appliances, sockets and plugs rated at 2 A can be wired to the lighting circuits; but a fused connector is used for electric clocks, with an outlet hole for the flexible lead. It is fixed by a central screw, since it is not intended for frequent withdrawal, and may be fused for 2 A or less; such a connector is shown in Fig. 70.

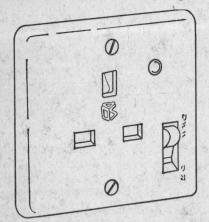

Fig. 68 Switched shuttered socket outlet with neon indicator light

(Dorman & Smith Ltd.)

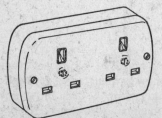

Fig. 69 Insulated surface-type twin 13 A socket outlets

(M. K. Electric Ltd.)

Fig. 70 Fused connector for electric clock (*M. K. Electric Ltd.*)

Fig. 71 Switched fused spur unit with indicator lamp and flex outlet (*M. K. Electric Ltd.*)

The fused spur box

This unit is used for taking a branch or 'spur' from a ring circuit for stationary appliances, and the fuse, which must not be more than 13 A, must not exceed the rating of the spur cable. It can be switched and provided with an indicator light, and can also have an outlet for a flexible lead in the front plate. It is suitable for serving a water heater or a fixed fire, where a socket outlet is not appropriate. A flush, insulated, switched fused spur box with neon light and flex outlet is shown in Fig. 71.

Lampholders

The lampholders employed for domestic lighting are of the bayonet type for tungsten filament lamps and should comply with B.S. 52. Two pins on the lamp cap go into slots, and the lamp contacts depress two spring-loaded plungers, with which they make contact. Small bayonet patterns (S.B.C. or B.15) are used for special purposes, including decorative schemes, but these should only be used where the circuit fuse is not larger than 5 A. Standard bayonet holders (B.C. or B.22) are used for all lamps up to 150 W. For lamp sizes larger than 150 W Edison screwholders are used (Fig. 72), and the lamp cap is provided with a screw thread, although 200 W lamps are also obtainable with bayonet caps. Although Edison screw lampholders are unusual in house installations, it is important to know that the centre contact must be connected to the line or phase wire of the circuit and the side or screw contact to the neutral wire.

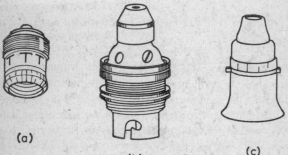

(a)

(b)

(c)

Fig. 72 (a) Edison screw lampholder
        (b) Metal bayonet cap lampholder
        (c) White moulded B.C. skirted lampholder

Brass lampholders that are in damp situations or that can be readily touched by a person standing on or in contact with earthed metal must be earthed; alternatively, insulated lampholders with protective shields that cover the lamp cap may be used. For domestic use 'all-insulated' lampholders made of moulded insulating material are commonly fitted. There is an outer shell of bakelite covering the metal reinforcement of the lampholder, which should be marked 'B.S. 52A'. This type of lampholder lasts fairly well with low-powered lamps, though the heating effect of 100 W lamps and greater powers may cause trouble with the shade ring, but heat-resisting types are made for temperatures greater than 135°C and are marked 'B.S. 52H'. These should be used with lamps in the 'cap-up' position in enclosed fittings or in fittings with inadequate ventilation. These lampholders are also shown in Fig. 72. In all lampholders provision should be made for gripping the flex, so that the lamp and shade are supported without the strain coming on the wires in the terminal sockets. Bakelite lampholders and fittings should be periodically examined to see that the bakelite is not cracked or chipped. If arcing or flashing occurs over the surface of bakelite, it leaves it in a dangerous condition, as carbon tracks may be formed. These give conducting paths for the current, and so its insulating properties are impaired and it is no longer safe.

Shielded B.C. lampholders are fitted with a 'skirt' or 'Home Office' shield of insulating material that covers the lamp cap and acts as the shade ring, which screws onto the bottom of the lampholder (as depicted in Fig. 72). Such shrouded lampholders are fitted in kitchens, sculleries, lavatories, bathrooms and cellars where there is a possibility of personal contact with damp walls or ground when replacing a lamp.

Adaptors

When it is required to connect more than one appliance to a single socket outlet, a fused adaptor plug is employed, which contains a fuse to protect the lower current circuit in the case of a 15 A adaptor with 15 A and 5 A outlets, or to protect the socket outlet from the connection of a greater load than its rated capacity in the case of a 13 A adaptor. Safe designs of adaptors are illustrated in Figs. 73 and 74; the cartridge fuse is accessible in one face of the 13 A adaptor shown, but in the 15 A adaptor the fuse is inside and the case has to be separated to replace it. 15 A and 5 A adaptors must comply with B.S. 546; smaller rated outlets than those

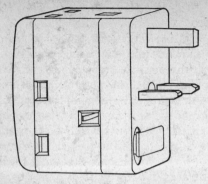

Fig. 73 13 A fused adaptor for B.S. 1363 sockets
(Nettle Accessories Ltd.)

Fig. 74. 15 A and 5 A fused adaptor complying with B.S. 546
(M. K. Electric Ltd.)

of the input pins must be fused on the input line; the adaptor must not be reversible; and all sockets must be connected to the same phase or pole as the corresponding input pins. Adaptors for use in 13 A sockets must also have the fuse in the input line, and the sockets and pins must correspond in the same way. There are several types of adaptors sold that do not comply with these requirements and are therefore unsafe to use.

Shaver supply units

The widespread use of electric shavers in bedrooms and bathrooms has necessitated a specially safe socket outlet to be devised for bathrooms, where shock risks are greatest and ordinary socket outlets and portable appliances are not permitted. Although electric shavers are generally of insulated construction, the shaver head is of steel and, since it makes very good contact with the skin of the face or neck, where shock could be painful and dangerous, a safer method of protection than that by earthing is adopted for bathrooms consisting of a double-wound transformer to isolate the shaver from the mains. This also enables a choice of different voltages to be offered through separate sockets on the shaver

Fig. 75 Shaver supply unit for bathrooms, flush type
(*Chilton Electric Products Ltd.*)

Fig. 76 Shaver supply unit for places other than bath rooms, surface type
(*M. K. Electric Ltd.*)

supply unit, which is convenient for foreign visitors to hotels,
The transformer core must be properly earthed, the shaver
unit must incorporrate a current-limiting device or a fuse not
exceeding 3 A and the unit must comply with B.S. 3052. For
use in places other than bathrooms, the transformer is not
essential, but the other requirements stated above are
necessary; the face plate should be marked FOR SHAVERS
ONLY, and if voltage selection is required an autotransformer
is permitted. A switched unit (which can be automatic) is
desirable for units with transformers to avoid overheating of
the windings when not in use. Shaver supply units are shown
in Figs. 75 and 76.

Switches and switching

A switch is a convenient piece of apparatus for opening or
closing an electrical circuit, and for most domestic circuits
tumbler switches or microgap switches are used. The single-
pole switch makes or interrupts the supply on one pole or
phase only; but when safety demands complete disconnec-
tion from the supply, double-pole switches are used.

Gas-filled lamps take about one-fifth of a second to attain
their operating temperature, during which time the current is
decreasing from about six times its steady value. With d.c.
supplies a quick 'make' of the switch contacts is necessary
so that heating and burning do not occur, and a quick
'break' is required so that the arc does not linger between
the switch arm and the contacts; a long break is also desirable to assist in extinction of the arc. But on a.c. supplies the
current is reduced to zero every half cycle, or 100 times per
second. Thus special requirements to prevent arcing are not
necessary, which is the reason why a.c. sockets and plugs
may be used to disconnect without switches. A very simple
switch mechanism is used in the a.c. switch by which the
current is interrupted by silver contacts separating at slow
speed, with a very small gap of 0·635 mm. This is called the

microgap a.c. switch; the mechanism is silent in operation, without buffers or other damping devices, and has a long operating life. The action is quick 'make' and slow 'break', but these switches must only be used with alternating current. They are noiseless because the only moving parts are the operating dolly and the flexing contact blade. The 'Mutac' switch is illustrated in Fig. 77.

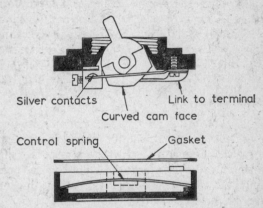

Fig. 77. Microgap switch   (*G.E.C. Ltd.*)

Types of switches for lighting

Surface-type switches with moulded covers are commonly used so that there is an insulating cover over the live parts. Sunk or flush switches with metal plates and dollies must have the metal parts earthed onto the conduit box, and earthed switches must be used. The standard size for lighting circuits is 5 A, though they seldom have to carry this current since, on 240 V, a 100 W lamp takes 0·42 A.

The 'underslung' type of switch is a good example of the d.c. switch, with the switch arm working like a pendulum in the porcelain base of the switch beneath the switch bracket. When breaking a circuit, the arc is drawn between the porcelain walls of the switch base remote from the cover.

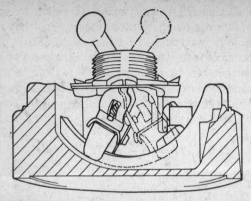

Fig. 78   Quick make and break tumbler switch

This is shown in the sectional arrangement of Fig. 78, but it is seldom used nowadays in domestic premises.

A popular type of tumbler switch made with either 'quick make and break' or 'silent toggle' action is illustrated in Fig. 79.

The latest type of a.c. switch with microgap action, which has attained a high degree of popularity, is known as the rocker switch. Instead of having a snap-action operating

Fig. 79   Surface-mounting switch with moulded dolly and cover
(J. A. Crabtree & Co. Ltd.)

dolly, a rocking-action lever works the switch by simply pressing in the projecting end of the rocker, which has only a small projection through the switch plate. This type of switch occupies such a small space that groups of two and three switches can be accommodated in a single square switch plate and box. They are narrow enough to use in the hollow spaces of metal door frames and are also used in 13 A switched socket assemblies. This type of switch is illustrated in Fig. 80.

Fig. 80 Group of three rocker switches

Suspension switches (or pear switches) are suspended by a flex from the ceiling. This type of switch should not be fitted when it can be avoided, since the supply voltage is on the conductors even when the switch is off and trouble is often experienced when the insulation of the flex becomes frayed and worn.

This type of switch is also incorporated in switched lampholders, and the same criticism applies if the lighting point is not switched elsewhere in the room. But where a switched lampholder on a lamp-standard is connected to a socket outlet, this is accepted as a convenient arrangement, although the flex should be inspected frequently for damage and wear.

Ceiling switches are preferable to the suspension switches previously mentioned. The ceiling switch is operated by a pendant cord, which is pulled either to switch 'on' or to switch 'off' consecutively. Such a switch is illustrated in Fig. 81. This type of switch is frequently used in bathrooms, where all switches must be out of reach of a person in the bath, and is useful in bedrooms so that the light can be controlled from the bed. Compared with a wall switch, a considerable amount of wiring can be saved, as the switch wires down the walls are dispensed with. This is shown in

Fig. 81 Ceiling switch

Figs. 82 and 83, in which the circuits for wall switches and ceiling switches are compared.

Push-button switches with consecutive action are often fitted to table lamps and suspension switches, and have a single projecting button which is pressed for either 'on' or 'off' as required. Push-button switches with reverse action are sometimes fitted to deep cupboards so that, when the door is opened, an internal light is switched on; closing the door switches the light off.

Switching circuits
One-way, single-pole switching has been illustrated when looping-in was considered, and various other switching arrangements will now be discussed.

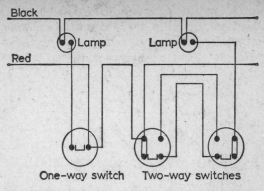

Fig. 82   Typical wiring with drops to wall switches

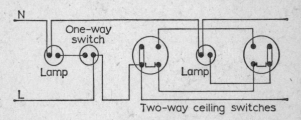

Fig. 83   Economy in wiring with ceiling switches

The circuit diagram illustrated in Fig. 84 shows a one-way switch with an additional 'loop-in' terminal for use where the cable run passes through the switch position. This avoids fitting a connector in the outlet box, which is bad practice. This type of switch is of interest because it provides a looping-out point for another point or an extension later.

All switches with metal parts that can be touched or are in contact with metal switch plates must be earthed, and outlet boxes or enclosures must be provided with an earthing terminal for this purpose, even if an insulated switch is fitted in the first instance. With earthed switches, the bridge of the switch is earthed. On one side of the switch bridge is an

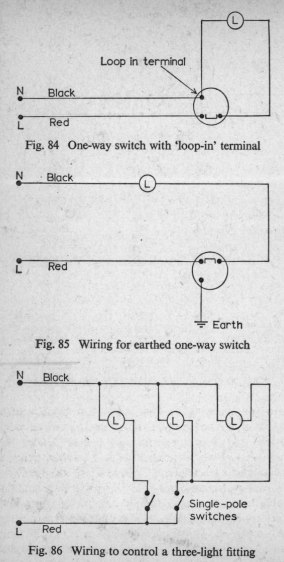

Fig. 84   One-way switch with 'loop-in' terminal

Fig. 85   Wiring for earthed one-way switch

Fig. 86   Wiring to control a three-light fitting

earthing terminal to take an earth wire, while on the other side is a brass lug which goes under the fixing screw when mounted on an earthed box, as shown in the Fig. 85. Where metal switch plates are employed, this feature also earths the plate.

Fig. 86 shows three lamps, of which one, two or three can be in use at any time by using two switches.

The double-pole switch is shown in the circuit of Fig. 87. Where neither pole of the electricity supply is earthed, linked double-pole switches must be used throughout a two-wire installation and in other installations for all heating appliances in which heating elements can be touched and which are not connected by means of a plug and socket. Thereby the danger of a single-pole switch in the black wire cannot occur, as the apparatus is entirely disconnected from the supply when the switch is 'off'. Double-pole switches can also be used for simultaneous control of two separate circuits.

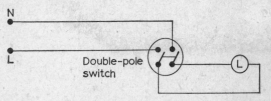

Fig. 87 Connections for a double-pole switch

Two-way switching is a convenience that is not generally appreciated; it is very useful in bedrooms, on the staircase, in corridors and in rooms with two doors. Besides being convenient, two-way switches save time and electricity, and avoid accidents in the house, most of which are due to falls. A two-way switch is a single-pole changeover switch, and it is possible to waste a considerable amount of cable by incorrect connection. The proper method of wiring is illustrated in Fig. 88. This is most economical on cable and is generally employed. Fig. 89 shows an unnecessary wasteful

method, and Fig. 90 shows the wrong method of interconnecting two-way switches in which both live and neutral
leads are brought into the switches. The risk of a short-
circuit is ever present, and if the cover is removed by anyone
it is dangerous. Regulations prohibit such an arrangement

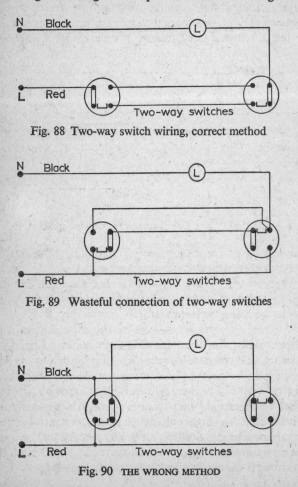

Fig. 88 Two-way switch wiring, correct method

Fig. 89 Wasteful connection of two-way switches

Fig. 90 THE WRONG METHOD

by not permitting a single-pole switch in the neutral line where this pole of the supply is earthed, but this method is illustrated in order to emphasise its danger.

Intermediate switching. Long halls, corridors and passage-ways with many doors call for switching the same group of lights off and on from more than two positions. This is done by using intermediate switches. A circuit embodying one

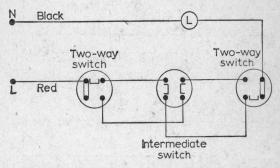

Fig. 91   Circuit with one intermediate switch

intermediate switch is shown in Fig. 91, but any number of intermediate switches can be connected in the same circuit by interposing them in the two 'strapping' wires between the two-way switches and connecting them in exactly the same way as the first intermediate switch shown.

Master control switching. This does not require a special type of switch and standard switches are employed. The master switch is so connected to the circuit as to have overriding control over the subsidiary switches. Fig. 92 shows a lighting circuit with each light individually controlled by a single-pole switch but with a single-pole switch in master control of the circuit as a whole. A double-pole linked switch could be used as a master switch so as to isolate both red and black leads of the circuit if the need was considered great enough.

There are many other switching variations for special purposes, such as dimming control and restrictive lighting, but any good installation contractor can devise special circuits with suitable switches for special applications.

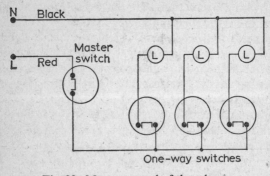

Fig. 92 Master control of three lamps

Dimmers and time switches

A new dimmer switch for room lighting is now available, but its adoption for domestic installations is slow as it is fairly costly. However, it has great advantages in dimming room lighting to a suitable low intensity for television viewing and for night lighting in nurseries or hospital wards. These units use electronic circuitry to reduce the voltage of the circuit, which is extremely effective and economical for lighting purposes, and some makes of dimmer switch can be accommodated in existing switch boxes to take the place of a switch. For television viewing, dimming is a great advantage because it avoids making the T.V. screen too bright in order to compete with normal room lighting, which is not good for the eyes or the T.V. set.

Time switches are being increasingly used in domestic a.c. installations nowadays. This type of control consists of an electric synchronous clock, which operates contacts 'on' and 'off' at predetermined times as set on an indicator dial. The most important use is for timing off-peak heating installations; and other minor applications are for time control of electric blankets, tea-makers, etc., for which plug-in-type units are available for use in socket outlets without any special wiring connections. Time switches are also incorporated in electric cookers for oven control.

In all cases of switch control, whether manual for lighting, heating or power, or automatic for time switching, it is always important to determine and specify the current rating required in each case, because this can range from the small 5 A lighting switch to the large 60 A or 100 A capacity control for a heating installation and too small a switch will overheat, burn at the contacts and break down very soon. The largest capacity controls usually have a high-capacity, solenoid-operated contactor switch controlled by a low-rated time switch.

6 Lighting

The evolution of artificial lighting has been a slow process, and for thousands of years outdoor visibility of large objects and daylight illumination were the normal seeing conditions. With the use of houses came some internal illumination during the hours of darkness, but this was of very poor quality at first, and reading and other close work were difficult. In more recent times the eye has been called on to a greater extent, and in the century following the Industrial Revolution the illumination provided was often inadequate for the work in hand, whether at the factory or in the home. Legislation has specified minimum standards of factory lighting and, with the need for maximum production, great improvements have ben made. The E.D.A. Division of the Electricity Council and the Illuminating Engineering Society have published much information on the proper use of electric light for different purposes. The latter body publishes recommendations for lighting building interiors with tables of illumination intensities for various kinds of work, known as the I.E.S. Code, from time to time as trends towards better lighting raise the standards of lighting. The values given in this chapter are taken from the I.E.S. Code.

Electric lamps

The earliest electric lamp for use in houses consisted of a carbon filament in an evacuated bulb. This was followed by the more efficient metal filament vacuum lamp and later by the gas-filled lamp. Modern lamps have tungsten filaments, but the bulb contains an inert gas consisting of nitrogen and argon. This enables the filament to be run at a higher temperature and gives greatly increased efficiency compared with the previous lamps for 40 W and above. Even so, the light

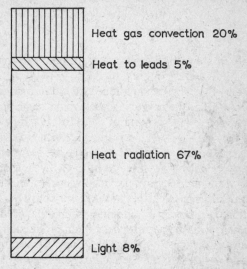

Heat gas convection 20%

Heat to leads 5%

Heat radiation 67%

Light 8%

Fig. 93   Energy distribution in gas-filled lamps

energy is a small fraction, about 8 %, of the total energy, as shown in Fig. 93. With gas-filled lamps the filament is in the form of a fine coil of wire less than a thousandth of an inch in diameter. The coil is formed in a close spiral to minimise the cooling effect of the gas. The latest filament

Fig. 94   Coiled-coil filament

development is in the 'coiled-coil' lamp, which is more efficient. This is due to a greater concentration of incandescent metal, which emits more light for the same operating temperature as a single-coil lamp. The filament arrangement is illustrated in Fig. 94.

Table 10 below compares the lumens output of coiled-coil and single-coil lamps and gives the increase of light with coiled-coil lamps.

Type	Electric power in watts		
	40	60	100
Coiled-coil: lumens	390	665	1260
Single-coil: lumens	325	575	1160
Coiled-coil: % increase in light	20	16	11

Table 10 Lumens output of coiled-coil and single-coil tungsten-filament lamps at 240 V

Light units

Lumen

A lumen (lm) is the unit of light flux used in describing the total quantity of light emitted by a light source or received by a surface.

Lux

This is the metric unit of illumination value and is equal to one lumen distributed over one square metre (0·093 lumen per square foot).

Candle-power—candela

The candle-power of a source is an old-fashioned term designating luminous intensity, which is now expressed in candelas (cd); a light source of 1 candela emits 4π lumens (12·57 lm).

Mean spherical illumination

This is expressed in lux but with a somewhat different meaning to that given above. It is the average illumination over the internal surface of a small sphere centred at the point of the light source, i.e. it is the incident flux on the surface of the sphere divided by the area of the sphere.

Illumination at a point

The illumination received at any point is inversely proportional to the square of the distance of the light source (Inverse Square Law) and proportional to the cosine of the angle of incidence. This is illustrated in Fig. 95, in which a lamp with a light intensity of I candelas in the direction shown is suspended h metres above the horizontal plane and d metres from the lamp to the point P. The illumination (E) at point P is given by:

$$E = \frac{I \times \text{Cos A}}{d^2} \text{ lumens/m}^2 \text{, or lux.}$$

As $\text{Cos A} = \dfrac{h}{d}$, $E = I \times \dfrac{h}{d^3}$ lux.

The lux or lumens per square metre is the unit of illumination generally used in room lighting calculations and is employed for measuring the lighting obtained. Common

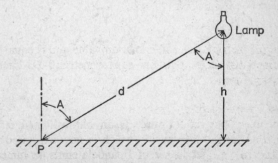

Fig. 95   Illumination from a lamp

values of illumination range from 0·0027 lx for 'starlight' to
such values for interior lighting in domestic premises as 50
lux for passageways and 600 lux for fine work like sewing and
darning. General illumination should not be less than 100 lx,
while visual tasks such as office work and sustained reading
should have 200–400 lx for adequate seeing conditions.

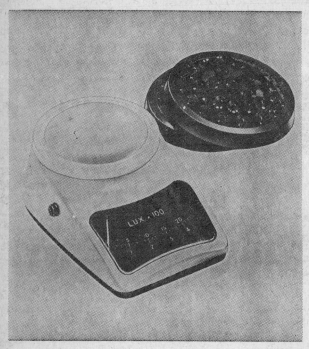

Fig. 96   Lightmeter, which gives illumination in lux, with muti-
plier attachments for the higher scale

(Sangamo Weston Ltd.)

Light measurement

The human eye is not accurate enough to measure illumination, though it can compare equality of brightness with fair

accuracy in photometers, in which the light is compared with a standard source.

Portable photometers or lightmeters are now extensively used to give direct readings in lux. Their action depends upon the effect of the light on a photoelectric cell; this generates a very small current which gives a reading on a milliammeter calibrated in lux. Such a lightmeter is illustrated in Fig. 96, and by its use measurements can be taken to check any lighting system when installed or at a later date to show the effect of dirt and ageing of the lamps. Before the introduction of international standards, lightmeters were calibrated in foot-candles. The foot-candle was the unit of illumination, which was 1 lumen distributed over a surface area of 1 square foot, and it is still used in the U.S.A.

Types of incandescent lamps

The bulbs are made either of clear glass or with the interior frosted (pearl). The clear lamp should not be used where there is possibility of glare, which will cause eyestrain. Clear-glass lamps should be enclosed in suitable shades or diffusing fittings. Pearl lamps give the same light output as clear lamps, but the diffusion obtained from the frosted glass lessens the glare and softens the shadows, particularly at lower mounting heights. Opal lamps have a thin skin of opal glass on the outside of the clear-glass bulb, but this type of lamp has been largely superseded by the white, internally-coated bulb. The diffusion is complete and shadows are minimised.

Daylight-blue lamps have the bulb made of a blue-tinted glass. The glass absorbs some 50% of the total light, so about double the wattage is often needed. Such lamps are better for colour matching, but the light is sometimes considered cold and depressing. Coloured lamps are available whose bulbs are internally coated blue, green, white, amber, red, pink or yellow. These all absorb high proportions of the light output—according to colour—of the filament, which makes them very inefficient for lighting purposes. The mushroom-shaped lamp has attained much popularity due to its attractive shape and smaller dimensions than the general-purpose lamp. It is internally coated white and provides an evenly diffused light.

A number of decorative types of lamp are also available in various shapes: small, round bulb; plain and twisted candle-flame shape; and long, tubular candle shape. These are suitable for some styles of decorative lighting fittings, but the lamps are of the inefficient, low-wattage, vacuum type. So-called 'long-life' lamps are simply under-run lamps, i.e. rated for a slightly lower voltage than the standard lamp and with reduced light output.

The incandescent filament lamp is relatively inefficient and gives from 8 to 14 lumens per watt, depending upon its size, for the normal domestic range of lamp wattages; and it has a rated life of 1000 hours. But it is cheap, easy to install and can be conveniently controlled by simple switches.

Electric-discharge lamps

An electric discharge through gases gives a much greater light efficiency than can be obtained with incandescent lamps. Discharge lamps are best suited and usually designed for operation on a.c. supplies, but with suitable control gear they can also be used with d.c.

Mercury and sodium electric-discharge lamps are used for street lighting and factory installations, but the single colour is a disadvantage, making them unsuitable for domestic and commercial uses. To overcome this the mercury discharge lamp bulb has been coated internally with phosphor, which converts ultra-violet radiation produced by the discharge into visible red light, so improving the colour rendering and producing a cold, bluish-white light which has been much preferred for street and factory lighting. These lamps are mostly produced in the larger sizes for this purpose, and although 50 W and 80 W lamps are obtainable the light is not considered 'warm' enough for home use. The efficiency is about three times higher than that of the gas-filled tungsten lamp and the rated life is 5000 hours.

The latest development in sodium discharge lamps is the high-pressure sodium lamp which is made to give a warm golden light and is becoming most popular for street lighting due to its higher efficiency, longer life and attractive colour rendering.

The tubular fluorescent lamp has been developed in a range of warm and cool colours, and, though more expensive initially than the tungsten lamp, the much higher light output obtained proves economical if the lamp is in use for long periods. The average light output for the first 5000 hours is 60 lm/W for 'warm white' and 57 lm/W for 'daylight' colours of a 1500 mm (5 ft) lamp rated at 80 W. The corresponding figures for a 1200 mm (4 ft), 40 W tube are 65 and 62 lm/W. Some colour variations are obtained by a coating of different phosphors and filter coats on the inside of the tube, but this reduces the light output. Rated life for the larger tubes in 7500 hours.

Some auxiliary apparatus is required, as shown in the circuit diagram of Fig. 97. The automatic starting switch may be of the 'glow' or 'thermal' type; the small condenser across it is to suppress radio interference. The starting device heats the lamp electrode filaments, and then opens and initiates the discharge at a high voltage from the choke, but the final running voltage across the lamp is about 110 V. The choke, on a.c. supplies, limits the current to the correct value but lowers the power factor, so a mains capacitor is fitted in parallel across the supply to improve the power factor. The 80 W lamp requires about 7·5 µF (microfarads) and the 40 W lamp about 3·2 µF capacitors. These starter circuits take several seconds to strike the arc. 'Instant start' lamps are available which have different control gears and circuits, and an 'earthing strip' along the length of the lamp which is necessary for reliable starting. Failure of the lamp is usually due to deterioration of the fluorescent powders in the lamp coating and evaporation of the electrodes; starter switches can be replaced, but failure is accelerated by fre-

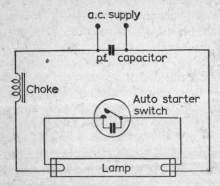

Fig. 97 Fluorescent lamp with glow starting switch

quent switching, and insufficient preheating prevents the lamp from starting.

The electric discharge between the electrodes in the tube goes out 100 times a second on a 50-cycle supply. This causes a stroboscopic effect when the light is reflected on rotating or

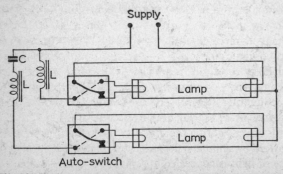

Fig. 98 Twin lamp circuits; eliminates stroboscopic flicker

moving objects which in certain cases is objectionable. To overcome this a twin lamp circuit can be used on a single-phase supply; the circuit is shown in Fig. 98. The upper lamp goes out at a different time in the cycle from the lower lamp, due to the different circuit constants, which puts their operation out of phase with each other, and the overall power factor is corrected automatically to nearly unity by this circuit.

Philips Electrical Ltd. produced a fluorescent fitting with a tungsten lamp acting as the choke for the fluorescent lamp. This saves cost and weight in the fitting, and the combination produces its own colour rendering with slightly lower efficiency; but the overall life is less than normal fluorescent lighting.

Fluorescent tubes are also made in circular form, 406 mm (40 W), 305 mm (32 W) and 209 mm (22 W) diameter, which are convenient for conventionally shaped light fitting designs.

Domestic lighting

There are three main requirements: sufficient uniform illumination, freedom from glare due to unscreened lights or reflection from polished surfaces, and freedom from deep shadows and abrupt contrasts. The light should not flicker or vary in intensity, but this is not apparent on 50-cycle systems with metal-filament lamps and is usually only evident as stroboscopic effect with fluorescent lamps if the twin lamp circuit described earlier is not used. The choice of fittings and shades depends on personal tastes and whether direct or indirect lighting is employed. The colour and surface of the walls have direct bearing on the effect obtained.

With direct lighting the majority of the light is projected downward by suitable reflectors, while in indirect lighting the source of light is concealed and illumination is obtained by reflection from ceilings, walls, curtains or other surfaces.

Indirect lighting is not used to any great extent in houses, except for decorative effects in addition to other methods of direct lighting. Semi-indirect lighting is employed with translucent bowl fittings and other types of fittings with opaque bottom panels and translucent side panels which considerably reduce the glare associated with direct lighting, especially when the shade employed gives insufficient cut-off.

With semi-indirect lighting, such as that with a thick diffusing bowl, or direct lighting with large silk shades or other inefficient types of fittings, additional socket outlets should be provided for desk or table lamps and lamp standards. In sitting rooms where reading is done, at the dinner table and at dressing tables not less than 200 lx should be available. In kitchens, wash-houses and bathrooms totally enclosed diffusing fittings should be used so that the lamp is protected against steam and dirt, and the exterior of the globes can be kept clean. Staircases and passages should be sufficiently illuminated in order that there is not too much contrast in coming out of a well-lighted room. An ample number of socket outlets should be put in at the time of installation, as they are essential for the multiplicity of portable electric equipment found in most homes nowadays, as well as for lighting standards. Multiple plug adaptors with trailing flexible leads are unsightly, and extra socket outlets provide a more convenient arrangement and a safer installation.

Colour
Artificial lighting contains different proportions of spectrum colours from daylight. Metal-filament lamps contain a larger proportion of red and yellow rays, and are deficient in green and blue. Colours in which red and yellow predominate appear much warmer under artificial light than in daylight, while green and blue surfaces appear dull because they absorb much of the light. The idea that green is a restful colour is only true when ample illumination is provided, since

eyestrain may result if too much light is absorbed. With dark colours considerably more light is necessary for good visibility than for light colours, whether referring to objects being inspected or to the surrounding walls and ceiling of a room which contribute their quota of light by reflection.

The ratio of reflection under a tungsten lamp to that with daylight, for different-coloured papers, is given in Table 11 below.

Colour of paper	White diffusing	Red	Orange	Yellow	Green	Deep blue
Ratio: $\dfrac{\text{Reflection under tungsten lamp}}{\text{Reflection with daylight}} =$	1·0	1·48	1·26	1·08	0·75	0·69

Table 11 Reflection ratios of tungsten lamp to daylight

The amount of reflection also depends on the surface texture. A glossy enamel finish will reflect more light than a flat tone or matt surface. Glossy surfaces give specular or directional reflection and matt surfaces give diffuse reflection.

Good mirrors and best white surfaces reflect about 90% of the incident light, and some approximate reflection factors are given in Table 12 below.

Material (clean)	Approx. reflection factor %	Material (clean)	Approx. reflection factor %
White tile, glossy	80	Ivory, matt	64
White paint, glossy	78	Light stone	58
		Middle stone	37
Plaster, matt white	70	Yellow brick	35
		Red oak	32
Ivory, glossy	69	Red brick	25

Table 12 Reflection factors

Shadow

The perception of an object in three dimensions is due to the light and shade effect, as well as to colour difference. There should be sufficient contrast of brightness between the object and its surroundings for easy vision, and the shadows must not be dense and black but soft and grey. Heavy shadows are dangerous as they are liable to cause accidents, especially on stairs and in passages. Direct lighting gives more pronounced shadows, while with indirect lighting there is an absence of hard shadows. Semi-indirect lighting combines both effects, with soft shadows from a bright ceiling and a degree of harder shadows from the light fitting.

Illumination calculations

In order to calculate approximately the size of lamp necessary to provide the required value of illumination in a room, the following formula can be used:

$$\text{Lumens per lamp} = \frac{\text{Average light flux required (lx)} \times \text{Area per lamp (m}^2)}{\text{Utilisation factor} \times \text{Maintenance factor}}.$$

Utilisation factor

This is the ratio of the total flux received on the working plane to the total flux produced by all the lamps in the room. Sometimes called the coefficient of utilisation, its value varies from about 0·3 to 0·5 with direct lighting in average conditions and rooms in houses and depends on the type of fitting used, the dimensions of the room, and the colour and condition of the walls and ceiling.

The utilisation factor is obtained from tables giving the classification of various types of lighting fittings, according to their downward light distribution, and by applying factors depending on room size (room index) and reflectances. This is a complicated process and can be studied in more detailed works on illumination, particularly the technical reports of the Illuminating Engineering Society. But reasonably satisfactory illumination calculations can be done by the simpler method of using the tabulated values given in Table 14 for typical fittings shown in Fig. 99, such as were successfully used by the lighting industry until recent years when more complicated factors covering zonal classification of light fittings were introduced.

Room index

The room index is first found by working out the expression:

$$\frac{\text{Room length} \times \text{Room width}}{\text{Height above working plane (length} + \text{width)}}.$$

For example, if the height is 2 m, and the room is 6 m long and 5 m wide, the room index is:

$$\frac{6 \times 5}{2(6+5)} = 1 \cdot 36.$$

The dimensions of a room affect the utilisation efficiency, because in a small room a large proportion of the light is absorbed by the walls, whereas in a large room with a number of lights a much larger proportion of the light from the lamps falls directly onto the working plane.

Effect of walls and ceiling

The colour and surface of the walls have a considerable effect on the illumination on account of their reflecting power. The lighter the colour of the walls and ceiling, the more the light is reflected and, consequently, the higher the utilisation factor. Allowing a small proportion for deterioration, wallpaper colours can be arranged as below in three classes: light with 50% reflection or more, medium with about 30% reflection and dark with about 10% reflection. The colours are arranged in order of reflecting power, and this should be taken into consideration when deciding the total light required.

Light: white, cream, yellow, light orange, light stone, light buff and pale tints.

Medium: grey, pink, light green, sky blue, stone.

Dark: dark grey, brown, red, dark green, blue.

Light fitting (Tu = tungsten lamp) (Fl = fluorescent lamp)	Ceiling Walls	Fairly light 50%		Very light 70%	
		Fairly dark 30%	Light 50%	Fairly dark 30%	Light 50%
	Room index	Utilisation factor			
(a) Open opaque reflectors; 75% light downwards (Fl or Tu)	0·6	·31	·35	·31	·36
	0·8	·4	·44	·4	·45
	1·0	·44	·49	·45	·49
	1·25	·49	·53	·49	·55
	1·5	·53	·57	·54	·58
(b) *Bare lamp on ceiling batten; 65% light downwards (Fl)	0·6	·22	·27	·24	·29
	0·8	·3	·35	·31	·37
	1·0	·35	·4	·37	·44
	1·25	·4	·45	·42	·49
	1·5	·44	·5	·47	·54
(c) Diffuser with open top and louvred beneath; 30% light downwards (Tu)	0·6	·2	·24	·23	·28
	0·8	·26	·3	·3	·35
	1·0	·3	·34	·34	·4
	1·25	·33	·38	·39	·45
	1·5	·36	·41	·44	·49
(d) Enclosed diffusers spherical or near spherical; 45% light downwards (Tu)	0·6	·16	·2	·18	·23
	0·8	·22	·27	·24	·3
	1·0	·26	·31	·29	·36
	1·25	·3	·35	·34	·41
	1·5	·34	·39	·39	·45
(e) Complete luminous ceiling of translucent corrugated strip or pan-shaped panels (Fl)	0·6	·15	·2	(ceiling cavity should be white and cavity depth not more than a third of the width)	
	0·8	·25	·28		
	1·0	·31	·34		
	1·25	·34	·37		
	1·5	·36	·4		

* The addition of a diffuser reduces the downward light to about 50% and reduces the utilisation factors by about 10%.

Table 13 Utilisation factors

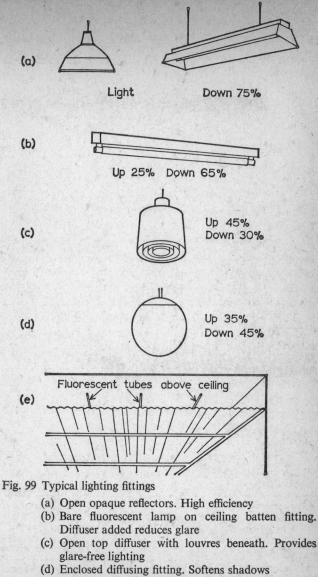

Light   Down 75%

Up 25%   Down 65%

Up 45%
Down 30%

Up 35%
Down 45%

Fluorescent tubes above ceiling

Fig. 99 Typical lighting fittings

    (a) Open opaque reflectors. High efficiency
    (b) Bare fluorescent lamp on ceiling batten fitting.
        Diffuser added reduces glare
    (c) Open top diffuser with louvres beneath. Provides
        glare-free lighting
    (d) Enclosed diffusing fitting. Softens shadows
    (e) Luminous ceiling with translucent strips or panels.
        Very even shadow-free lighting

Ceilings can be divided into very light—generally white or cream and very pale tints—with 70% reflection, fairly light with 50% reflection (as for light walls above) and dark with 30% reflection (as for medium walls above). The effect of smoke and dirt will lessen these figures.

Maintenance factor. This allows for the falling off of efficiency due to deterioration of reflectors, walls and ceilings caused by dirt. With cleaning every few weeks, the average illumination obtained will be about 80% that of the original clean and new conditions, and a maintenance factor of 0·8 is usually taken for normally good conditions. This can be increased for extra-clean conditions but can fall to as low as 0·4 or 0·5 in dirty industrial surroundings.

Table 13 gives typical basic forms of light fittings illustrated in Fig. 99, together with the related utilisation factors for various room index values. These five sections and figures should meet most domestic applications.

The typical lighting fittings illustrated are simple diagrams, but some more decorative domestic shades may give results that differ from the calculated values.

The lumens obtained from 240 V tungsten lamps, ranging from 40 W to 100 W with coiled-coil filaments and from 150

Tungsten		Fluorescent		
Watts	240 V	Watts	Warm white	De luxe warm white
40 (pearl)	390 (c.c)	20 (2 ft)	1100	750
60 "	665 "	30 (3 ft)	1900	1350
100 "	1260 "	40 (2 ft)	1700	1100
150 "	1960 (s.c)	40 (4 ft)	2600	1950
200 (clear)	2720 "	65 (5 ft)	4400	3000
300 "	4300 "	80 (5 ft)	4800	3400

Table 14 Nominal average lumens output throughout life

Note: 110 V lamps have approximately 11½% greater light output than 240 V lamps.

W to 300 W with single-coil filaments, and from fluorescent
lamps, ranging from 20 W to 80 W, in white, warm white and
de luxe warm white types (these lamps being commonly
used in domestic premises) are given in Table 14.

To ascertain lamp size
From the foregoing the lamp size for any room can be
worked out, and an example will now show the application of
this method.

Example 13. A sitting room 3 m by 4 m has a single lamp,
open top, diffuser fitting with louvred base for a tungsten
lamp which gives direct illumination and is fitted 2 m above
the working plane. The walls are light and the ceiling is very
light. Assume a depreciation factor of 0·8. What size lamp is
required for a general illumination of 100 lx?

Area per lamp is 12 m². The room index is $\dfrac{3 \times 4}{2(3+4)} = 0.86$.

From Table 13 (iii), starting at room index between 0·8 and
1·0, the fourth column of factors gives between 0·35 and 0·4,
say 0·36, as the utilisation factor.

$$\text{Lumens per lamp} = \frac{100 \times 12}{0.36 \times 0.8} = 4166.$$

From Table 14 a 300 W lamp gives 4300 lm, and this is the
correct lamp to use. However, being somewhat larger than
necessary, it will provide approximately 3% more light, or
about 103 lx, which is near enough to requirements. This
being the average value, illumination will be much higher
under the light fitting and lower towards the outer parts of
the room. For dining rooms and sitting rooms the average
value of illumination should be 100 lx, so the above value is
not excessive; but further local lighting is desirable for reading and sewing, which require 200 lx and 600 lx respectively.
With larger rooms, especially long rooms, more than one

lighting fitting is necessary for better light distribution, unless lamp-standards or wall brackets are provided, but wall fittings are often more ornamental than useful.

In drawing rooms the average value should be about 200 lx, but sufficient outlets should be provided for the use of standard or table lamps, and a piano should be so illuminated that the player does not cast a shadow on the music.

Bedrooms require about 50 lx generally, with 200 lx at the bedhead for reading. One light should be over the dressing table, and additional outlets should be provided for bed lights and fitted basins. A separate central light for the wardrobe is often useful, especially if it contains a long mirror. Two-way switching should be employed so that the main light can be switched on at the door and off from the bedside. For the latter switch, one of the ceiling type is safer than a pendant switch.

Bathrooms, toilets and washrooms should have protected fittings, either of the enclosed type or employing insulated lampholders with skirts. In rooms containing a bath or shower, switches must not be within reach of the person using the bath or shower. Therefore, if the normal switch position by door is not out of reach, the switch must either be a cord-operated ceiling switch or be fitted outside the door. Socket outlets must not be provided in bathrooms, but a shaver supply unit complying with B.S. 3052 is permitted.

Kitchens require at least 200 lx by modern standards, and it is very desirable to increase this to 400 lx at the cooker, sink and work table.

Example 14. A kitchen, 3 metres square, is provided with two enclosed diffusing fittings mounted on the ceiling, which is 2 m above table level. Each fitting contains a 240 V, 60 W tungsten lamp, giving 665 lm. The ceiling and walls are of a light colour, and the maintenance factor is 0·8. Estimate the illumination provided by the original installation and recommend modern requirements.

The layout of the kitchen is given in Fig. 100. The two lights are arranged diagonally, so that the cooker, sink and fitted cabinet with its work table are illuminated. Switches are provided at each door, as shown on the plan, each controlling one lamp for convenience in switching on the light whichever entrance is used.

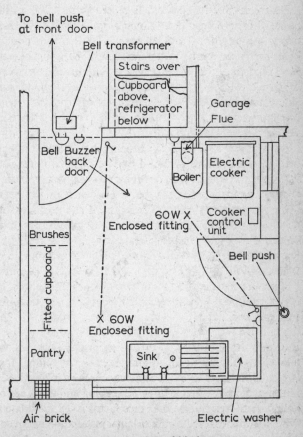

Fig. 100 Layout of kitchen

The room index is $\dfrac{3 \times 3}{2(3+3)} = 0.75$, and the utilisation factor is 0·28, deduced from section (d) last column, Table 13, page 138.

$$\text{Area/lamp} = \frac{3 \times 3}{2} = 4.5 \text{ m}^2.$$

The illumination (lux) =
$$\frac{\text{Utilisation factor} \times \text{lumens/lamp} \times \text{Maintenance factor}}{\text{Area/lamp}} =$$
$$\frac{0.28 \times 665 \times 0.8}{4.5} = 33 \text{ lx}.$$

This figure is too low by modern standards, and a 200 W or 300 W tungsten lamp should be used in each fitting. These would bring the illumination up to 110 lx or 174 lx respectively. However, few modern households maintain the high standards recommended by the I.E.S. principally for commerce and industry, and most would probably find 150 W lamps, producing almost 80 lx, adequate. The kitchen is most suitable for the application of fluorescent lamps, and it will be seen that, by adjusting the utilisation factor to 0·35 for one fluorescent batten fitting in the centre of the ceiling with a bare 1500 mm (5 ft), 80 W warm white lamp, 122 lx would be produced; alternatively, two 1200 mm (4 ft), 40 W lamp fittings would produce 132 lx, which would be much more economical than using tungsten lamps.

Sculleries and wash-houses should have a light over or beside the sink, not behind the person using it. The switches should be out of reach of a person standing at the sink with wet hands, or they should be of the cord-operated ceiling type; although this is not a requirement of Regulations and the situation is not as dangerous as that which exists in a bathroom, the risk of a less serious shock is nevertheless still present. Enclosed fittings are best for damp situations, but in any case the lampholders should be properly earthed if of metal or, preferably, be of the insulated type with protective shields or skirts.

Staircases and halls should have from 50 lx to 100 lx, and the light should show the treads of the stairs and any changes of level.

7 Electric Heating

Although the open coal fire is used in living rooms on account of its cheerfulness, it is not generally appreciated that only some 25 % of the heat energy of the coal effectively warms the room.

Assuming that when 1 kg of coal is burnt 28 MJ are liberated, only about 7 MJ are radiated into the room.*

One kilowatt-hour or one unit of electricity is a constant quantity in all localities; the quantity of heat contained in one unit is 3·6 MJ, and the efficiency of conversion to heat is 100%. Therefore, all the heat given off by an electric heater is used to warm the room. The installation costs of an electric heating system are usually less than those of other systems; maintenance costs are much lower, and control, cleanliness and convenience are so much better. Thus, although electricity is the most expensive of fuels, its advantages in other directions make it extremely popular for heating purposes in the home. Low rates for off-peak heating also offer savings in running costs and help to make electricity competitive with other systems of heating.

Radiant and convector heaters

Space heating equipment can be divided into two main types: the radiant type, in which there is a red-hot element, with or without a reflector, such as the popular 'electric fire'; the non-luminous radiant panel type and the convector type of heater. The true convector transfers almost all its heat to the

* MJ = megajoule or 10^6 joules—the energy, in this case, in heat.

air current passing through it, whereas the radiator transfers most of its heat directly to the objects on which its radiation falls; this is about 70% in the case of the reflector fire heater. Types of heater that transmit heat both by convection and by radiation in varying proportions include the tubular heater, the hot water and oil-filled 'radiator', and the panel heater.

Fig. 101 Combined radiant and convector heater
(Belling & Co. Ltd.)

There are also combined radiators and convectors, which have the advantage of separately controlled radiant and convected heat. This type of combined heater is illustrated in Fig. 101. The appearance of warmth is obtained by an amber lamp in the base opening, through which the warm air is circulated. This model obtains five degrees of heat output from $\frac{1}{2}$ kW to $2\frac{1}{2}$ kW by means of the switches at the base. A popular heater is the fan heater, which combines a heating element and fan in a very small space to blow out a strong current of warm air for rapidly heating a room. These units can stand on a table or on the floor. However, the fan comprises a feature that can break down and reduce the reliability of the heater, in that the heater cannot operate safely without overheating if the fan is not working.

Fig. 102 2 kW reflector fire

(Belling & Co. Ltd.)

A modern-style heater providing a high proportion of direct radiated heat is illustrated in Fig. 102. Safety-guards must be fitted to all electric fires to comply with B.S. 1945 and with the Heating Appliances (Fireguards) Regulations 1953. Portable heaters are usually fitted with 2 m of three-core flexible cord.

A heater designed for wall mounting is shown in Fig. 103.

Such a fitting is suitable for the bathroom, but it must be mounted high up on the wall out of reach of a person using the bath.

Location of heater

It should be understood that a convector heater will only heat the air above its own level because the warmed air rises in the room. It must therefore be placed as near the floor as possible,

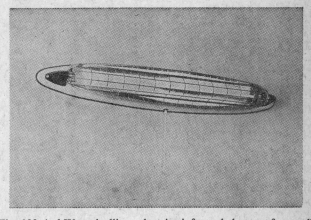

Fig. 103 1 kW swivelling electric infra-red heater for wall mounting
(Dimplex Ltd.)

otherwise a layer of cold air will always remain below the level of the heater. Since radiation is always emitted in straight lines in all directions from the element (as from the sun) and only redirected by a reflector, a radiant heater such as the infra-red type (so called) and other reflector 'fires' can be placed at any level desired, provided that the angle of the heater is directed towards the part of the room where the heat is mainly required.

It is best to locate convector heaters in the coldest parts of a room, which are generally below windows where the air, cooled by the cold glass, starts a cold down-draught. Radiant

Room length	Room width					
metres	2	2·5	3	3·5	4	5
2·5	$\frac{3}{4}$	$\frac{3}{4}$	1	1	2	2
3	$\frac{3}{4}$	1	1	2	2	$2\frac{1}{2}$
3·5	1	1	2	2	$2\frac{1}{2}$	$2\frac{1}{2}$
4	1	2	2	$2\frac{1}{2}$	$2\frac{1}{2}$	3
5	2	2	$2\frac{1}{2}$	$2\frac{1}{2}$	3	3
5·5	2	$2\frac{1}{2}$	$2\frac{1}{2}$	3	3	3

Table 15 Guide to heater sizes for various rooms (kW) (not off-peak storage heating)

Note: This table is for living rooms with average conditions and assuming a room height of 2·5 m. The loadings can be reduced by 20% for bedrooms, halls, etc., but must be increased for occasional occupation, abnormal exposure with cold walls or with large window areas.

heaters are best placed on or against the coldest walls. Heaters that depend on a large proportion of heat emission by convection from their surfaces should have plenty of air space around them, especially wall-mounted panel types, which should be spaced not less than 50 mm off the wall.

It is most important with all types of non-luminous heaters to ensure free heat dissipation at all times, because restricted ventilation causes the heater temperature to rise until either the heat can escape or a dangerous fire is started in the room. Therefore, clothing and the like should never be placed on such heaters to air or dry. This is a most dangerous practice which has led to catastrophic fires on a number of occasions. All non-luminous heaters should have a pilot lamp or lamps, either self-contained or situated somewhere in the room (in the thermostat or control switch) to indicate when the heater is operating.

Convector heaters run at a lower temperature than the elements of an electric 'fire', and thus their elements have a longer life and require less maintenance; they also constitute less of a fire-risk. Convector heaters are more suitable for continuous use for space heating, as it takes time for the warm air to be circulated in contrast to the immediate effects of radiant heating.

Temperature control

Thermostatic control should be employed to avoid wasting electricity when the desired temperature has been reached and to maintain a reasonably constant temperature, particularly where continuous or long-period heating is required, say for several hours. Thermostats can be provided separately as part of the wiring installation and fitted in each room, or they can be incorporated in convector-type heaters at low level to sample the cool air flowing into the heaters. In this way each heater has its own automatic controller. It should be appreciated that comfort can be obtained from a radiant heat source with an air temperature several degrees lower than would be necessary with convected heat—for example, in sunshine on a snow-capped mountain. Therefore, air thermostats controlling radiant heaters may not be so effective as when controlling convectors but will, nevertheless, serve the purpose reasonably well. Most room thermostats are either of the bimetal strip type, in which differential expansion of two dissimilar metals with rising air temperature causes contacts to open, or of the bellows type, in which the expansion of a liquid or vapour in a sensitive phial, with rising temperature, causes a bellows to operate the contacts; the phial may be remote and operate the thermostat through a length of capillary tubing. An important feature of thermostats is the temperature differential between switching on and off; this can range from $\frac{1}{4}$°C to 1°C or more, and it is obvious that the closer the differential, the less the variation of room temperature will be with the switching of the thermostat. An accelerator heater within the thermostat is often utilised to hasten the action and so reduce the differential. The switching action makes the current capacity of thermostats very limited with d.c. but ample with a.c. It is generally necessary with d.c. for the thermostat to operate a larger capacity contactor in the heating circuit, but with a.c. thermostats can handle the full circuit current up to 15 A or 20 A directly.

Internal and external thermostats can be combined with a time switch and contactor for installation control, and a number of manufacturers design and supply such control units. With thermostatic control the heat output is limited to that required by the heat losses and air temperature. Thus the electrical consumption will be the same whether the loading installed is just adequate or excessive, and a high heater loading does not involve excessive electricity consumption but can cope with wider temperatue differences. Fig. 104 shows a typical oil-filled panel radiator fitted with a thermostat. These heaters provide approximately half the heat output as radiation and half as convection. The heater element is immersed in an oil reservoir at the base and the oil circulates through channels in the panel. In modern buildings convector units can be concealed in the walls or in the window recess, so long as air inlet and outlet grilles are provided, thus presenting a smooth exterior surface. This type of heater usually employs finned tube units, the heating element being contained in the tubes.

Tubular heaters

This type of heater warms a room partly by radiation and partly by convection, in a similar way to the oil-filled panel heater; it consists of a hot wire element in a circular steel tube about 50 mm diameter or in an oval tube about 76 mm by 25 mm. The lengths of the tubes vary in standard sizes from 300 mm to 3·6 m and are supported by wall brackets or floor mountings. The heating element inside the tube consists of a nickel-chrome wire, which runs at black heat and is supported on mica insulators. One end of the tube is closed by a steel cap, while the other end carries a terminal assembly with two brass terminals connected to the element and a brass earthing terminal solidly connected to the tube. The standard

Fig. 104 1 kW oil-filled electric panel radiator

(Dimplex Ltd.)

loading is 200 watts per metre, but elements for 265 watts per metre are also made. With the lower loading the surface temperature of the tube is about 80°C, while with the higher loading the surface temperature is about 106°C. These temperatures are much too hot to touch and, unless the heaters are fixed in a safe place where they are not likely to be touched, they should be fitted with a suitable wire guard. Fig. 105 shows the arrangement of two-tier tubular heaters with connections to conduit under the floor and a thermostat in the corner of the room well away from the heaters. The

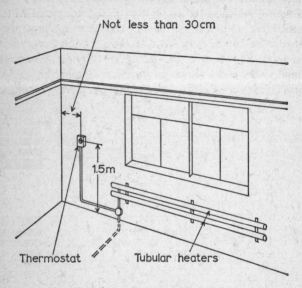

Fig. 105 Tubular heaters and thermostat

heaters should be situated around the skirting under windows to minimise the effect of down-draughts and cold-air inlets. This type of heater is often used at the base of high windows and skylights for this purpose. Although the tubular heater is much used for industrial, agricultural and other applications, it has been largely superseded for house warming by the less conspicuous skirting heater and other forms of heater, because it takes up too much wall space in a room and is not very attractive in appearance.

To calculate the heat required

The design of a heating installation with any degree of accuracy is a matter for expert and laborious calculation, as it involves working out the areas of room surfaces through which heat is lost and multiplying each surface area by the 'U' value, which is the amount of heat transmitted to the outside per unit area, per degree difference in temperature between inside and outside; multiplying by the temperature rise required, which is the difference between the room temperature and the lowest outside temperature to be considered; and adding together all the various heat losses to

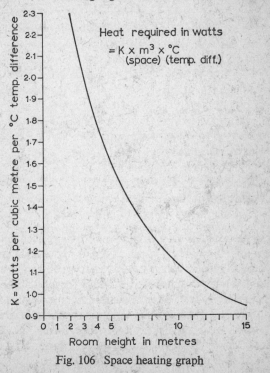

Heat required in watts

$$= K \times m^3 \times {}^\circ C$$
$$\text{(space)} \quad \text{(temp. diff.)}$$

Fig. 106 Space heating graph

obtain the continuous heat loss per hour under the worst conditions. The 'U' values are obtained from published tables for various building constructions under average normal conditions of exposure. The heat loss has then to be balanced by an equivalent heat input from the heaters.

There are, however, several simpler and approximate methods of ascertaining the heater loading required, based on average conditions and only involving a knowledge of the room dimensions. The graph given in Fig. 106 enables calculations to be made for normal buildings when the window glass area is not more than $0·6$ m^2 per 100 m^3. The height of the room being known, the watts per cubic metre of space can be obtained from the graph for a temperature difference of 1°C between the inside and the outside of the building, and a proportional figure will apply for any temperature differences. There are also tables published, such as Table 15, page 150, that give a reasonable guide to the loadings required where only very approximate results are needed.

The following example shows how the graph in Fig. 106 is used to ascertain the length and loading of tubular heaters that would be required to maintain the heat in a room for specified conditions.

Example 15. A small bedroom $3 \times 2·5 \times 3$ m is to be warmed to an inside temperature of 10°C, with the outside temperature 0°C. What length of tubular heater, run at 200 watts per lineal metre, would be required, and what would be the total cost of energy for 1000 hours per annum at 1p per unit?

Size of room	$= 3 \times 2·5 \times 3 = 22·5$ m^3.
Temperature difference	$= 10 - 0 = 10$°C.
From graph for 3 m height,	$K = 2$.
$\therefore$ watts	$= 2 \times 22·5 \times 10 = 450$.

At 200 watts per metre run, the length required is $2·25$ m, but if the nearest standard length made is $2·4$ m the loading will be 480 W.

Total number of kWh $= \dfrac{480}{1000} \times 1000 = 480.$

Total cost at 1p per unit $= 480 \times 1\text{p} = £4\cdot80.$

In practice, and for intermittent heating, a higher load-ing would be installed to raise the temperature from cold within a reasonable time.

Panel heaters and ceiling heating

There are two main types of radiant panel heating: with high-temperature panels, which work at a surface temperature of approximately 204°C; and with low-temperature panels, which operate at much lower temperatures, from about 93°C for wall panels to about 32°C for ceiling panels and 24°C for floor warming. The former type are usually fixed out of reach at high level and suspended or mounted on inclined brackets so that the radiant heat is directed down-wards. The latter type can be fixed against a wall or embedded in either the walls or the ceiling. Even with the lower temper-atures, the backs of the panels transfer a fair amount of heat to the structure, and heat-insulating pads are desirable on outside walls behind the panels. Similarly, roofs above top-floor rooms with ceiling panels should have extra good insulation to limit the heat losses increased by the heaters themselves.

Ceiling heating is usually designed for a maximum surface temperature of about 32°C so as to avoid discomfort with too intense heat on the head. For this reason, it is unsuitable for low ceilings, and the best results are obtained with ceiling heights over 2·5 m. The limiting surface temperature necessi-tates low values of watts per square metre, and therefore the whole ceiling area must be utilised to obtain the requisite load for the room. It is efficient, with good insulation. The heat response is fairly rapid and there are no obstructions in the path of the radiant heat; and the heat distribution is excellent. Thermostatic control should be employed with either type; the diversity of operation of a number of thermo-stats reduces the maximum power demand in large installa-tions, but in small installations, such as the average house, the main cables should be large enough to deal with the total current.

Floor heating and thermal storage

Electric floor heating has become popular in recent years because of its low initial cost compared with that of other forms of central heating and its high amenity and comfort values, but it has proved more costly to use unless a low cost per unit is available. Therefore, from an economical point of view, it is only suitable for continuous heating with off-peak supplies and tariffs in which low unit rates operate during restricted hours. It is admirably suited to this because the floors become storage heaters, storing the heat input during off-peak hours and emitting heat continously over the 24 hours.

This method of operation is quite suitable for houses generally, but in order to be as economical as possible a compromise scheme is usually adopted by which the ground-floor or living rooms are provided with floor heating to provide a limited temperature rise—say, to 12°C or 15°C— as 'background' heating, keeping these rooms mildly warm continuously, and other heaters on unrestricted supply (at higher tariff rate) are used for shorter periods and in bed-rooms as required to boost to higher comfort temperatures. The additional heat required for this purpose—say, for an extra 3°C or 5°C—is very small and is well covered by a 1 kW heater in most average rooms.

As with ceiling heating, the surface temperature is limited to about 24°C to avoid discomfort to the feet. The loading per square metre is therefore low, and the whole of the floor area is usually required to obtain the necessary electrical loading. The presence of furniture and floor coverings is not as a rule a very great disadvantage because, although they have an insulating effect and retard the emission of heat, the floor temperature in the covered area rises to overcome this until the required emission is produced. Floor heating is very effective for comfort conditions and, as with ceiling heating, no space is taken up by heating equipment, which is a great advantage. The heating elements can be embedded in the walls in addition to or instead of the floor. Floor or wall heating for unrestricted supply requires good thermal insulation to reduce heat losses behind the elements or panels, and wood-joist floors and partition walls can be used. The heat response is fairly rapid and very good heat distribution is obtained, but heating panels embedded in walls and ceilings require special heat-resistant paints and finishes if the heated areas or lines of the elements are not to show through in time. For off-peak heating, however, the thermal storage

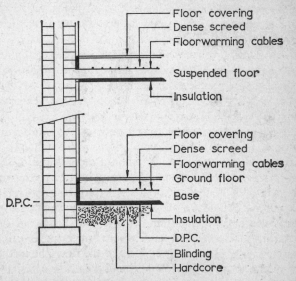

Fig. 107 Typical floor warming construction for off-peak heating installations

necessary requires a mass of heat-storage material; thus wood-joist floors cannot be used for this form of heating, which is usually only applied to solid concrete floors or to hollow tile floors with sufficient thickness (not less than 65 mm) of cement screed on top (see Fig. 107).

Thermostatic control for floor heating with unrestricted supply is, of course, essential, but with off-peak heating it is not possible to control or regulate the heat output of a heated floor artificially and natural regulation is the only means of control. In this way, the heat output depends on the difference in temperature between the floor surface and the air. Therefore, as the room temperature tends to fall with cold weather, the floor emits more heat to compensate for the heat loss, and when the temperature rises with warmer weather the heat emission from the floor is reduced accordingly. Thus the temperature conditions of the room and floor are constantly trying to balance; but the floor can never raise the temperature during the period of heat discharge in supply-restricted hours because it is constantly losing heat with falling temperature, and the temperature of the room gradually, though slightly, falls during this period.

Block storage heaters

The block storage heater was a natural development from floor heating for off-peak supplies, but it has the great disadvantage of size and weight, because its heat storage blocks must have enough bulk to take in sufficient heat during the few off-peak hours to escape gradually over the whole 24 hours in order to keep the room comfortably warm. This type of heater is extremely popular, and with the advent of the simple two-rate, off-peak tariff—generally known as the white meter tariff—in all areas it has enabled off-peak electric heating to be installed in homes at low initial cost and reasonable running costs. It is essentially a convector heater, and in its simplest form consists of a number of

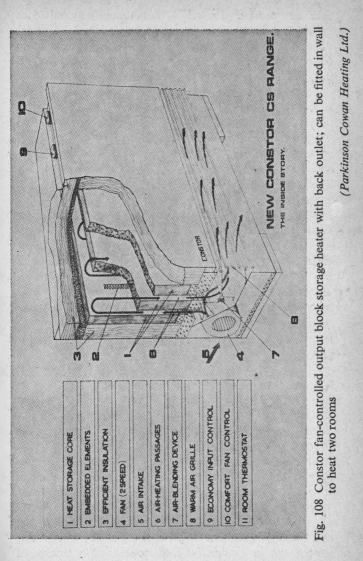

NEW CONSTOR CS RANGE.
THE INSIDE STORY.

CONSTOR

1 HEAT STORAGE CORE
2 EMBEDDED ELEMENTS
3 EFFICIENT INSULATION
4 FAN (2 SPEED)
5 AIR INTAKE
6 AIR-HEATING PASSAGES
7 AIR-BLENDING DEVICE
8 WARM AIR GRILLE
9 ECONOMY INPUT CONTROL
10 COMFORT FAN CONTROL
11 ROOM THERMOSTAT

Fig. 108 Constor fan-controlled output block storage heater with back outlet; can be fitted in wall to heat two rooms

(Parkinson Cowan Heating Ltd.)

blocks of cast iron or refractory material through which the heater element wires pass and to which the heat is transferred, so raising their temperature several hundred degrees. The blocks are surrounded by about 50 mm of suitable insulating material and are enclosed in a metal case with a fusible link-in circuit as a protection against overheating. In common with off-peak floor heating, this form of storage heater has no control over the heat output and thermostats cannot be used for this purpose. It is possible, however, to obtain this type of construction with air passages through the heater blocks and shutter regulators in the output air passage which can regulate the flow of air through the heater and the heat output to the room to a limited extent.

The best and most economical form of storage heater has a fan-controlled output. In this, a maximum of insulation is used to minimise the standing fixed heat loss from the heater, and a fan in the air passage forces air through the heated blocks to give a heat output that is under complete control, simply by switching the fan on or off as required. This is best accomplished automatically by a room thermostat and, provided that the heater has enough storage capacity, the room temperature can be closely controlled or varied at will throughout the day. Fig. 108 shows this type of heater with cut-away section to illustrate its action. These heaters are usually provided with adjustable internal thermostats to limit or adjust the amount of heat to be stored as required, and some have a time switch to run the fan at high speed for a short period for rapidly heating up the room when switched on in the mornings.

Calculating heat storage loading

The method of estimating the loading required in a storage heater is a simple calculation, since the total continuous heat output in watts for 24 hours has only to be divided by the number of off-peak hours of heat input.

Thus, assuming the maximum requirements for a particular room with a controlled output heater are 1·5 kW of heat output for 15 hours of occupation and 0·5 kW for the remaining 9 night hours each day, and the off-peak hours are 11.00 p.m. to 7.00 a.m , or 8 hours at night, then the loading required will be not less than

$$\frac{1\cdot5\times15+0\cdot5\times9}{8} = \frac{27}{8} = 3\cdot4 \text{ kW.}$$

The fixed (uncontrolled) standing heat output of these heaters is about 15% of the rating, so the standing heat output during the night while the fan is switched off would cover the night requirements in this case. It will be obvious that, if the off-peak tariff includes a period of 2 or 3 hours during the day when the heaters can be boosted, the longer charging period per day will proportionally reduce the loading of the heaters required.

Electricaire systems

Central heating for a number of rooms on the same principle is becoming a popular application of electric storage heating. These systems, known as Electricaire, employ a much larger version of the fan-controlled output heater, and the heated air is carried from a centrally placed heating unit through ducts to the various rooms in the house.

It is evident that, if thermal storage systems are not well worked out and properly designed, the user may find he has insufficient heat stored in the heaters for his maximum requirements and may wrongly blame the system instead of the designer.

For all forms of off-peak heating with off-peak tariff, separate circuit wiring for the heaters that is under the control of a time switch is necessary, but the fan circuits must be supplied from unrestricted supply circuits so that they may operate in conjunction with the thermostats during the day when the off-peak supply to the heaters is switched off. With the white meter tariff, however, which does not discriminate between different usages or heating equipment, such separate circuits are not essential, although in order to get the advantage of the low off-peak rate at night the heaters must be switched off or unplugged each morning for the peak hours. However, as this procedure is inconvenient and not altogether reliable, it is better to go to the expense of separate wiring and a time switch for a more satisfactory and trouble-free installation.

Position of thermostats

The function of a thermostat is to maintain an equable temperature as well as to conserve electrical energy. Thermostats should be freely exposed to air but not directly subjected to radiant heat or draughts. If thermostats are fitted close to the floor, they are in the paths of cold draughts, especially when situated near the side of the door that opens. Warm air increases in temperature towards the ceiling, and in some rooms with convection heating and an average temperature of 16°C it has been known to vary from 10°C at floor level to 20°C at ceiling level. So, if a thermostat is set at 18°C and is placed too low, then the mean temperature of the room at the higher planes occupied by the human body will be in excess of a comfortable working temperature. A suitable position is in a fairly protected place, preferably on an outside wall, about 1·5 m above floor level, as indicated in Fig. 105 (see page 154).

Electric blankets

The electric blanket consists of a spiral insulated element wound on a tough flexible core distributed over the blanket area and sandwiched between the blanket material. A switch with a neon lamp indicator is usually fitted in the flexible lead, and some types have thermostat contacts in the blanket itself to prevent overheating. The loadings are usually about 50 W for single bed size and 100 W for double bed size. Electric blankets should be inspected and tested, and any necessary repairs carried out, before use each winter period to ensure that they are in good condition. The dangers lie in unintentionally leaving them switched on during the day and in piling clothing on top of the bed while they are switched on, resulting in overheating and often causing fires. There are two types available: the underblanket and the overblanket. The electric blanket should never be used as an underblanket unless it is suitably protected by a thermostat or time switch; as an overblanket, it is less liable to become overheated.

Electric cookers

Electric cookers, apart from the single hotplate, vary in size from the small breakfast cooker to the heavy-duty cooking equipment found in canteens and hotels. This section, however, will be confined to domestic types of cookers, and the electrical and technical aspects of them.

The uniformity of the amount of heat provided from one unit (kWh) and the consistent results that can be obtained are one of the greatest advantages of the electric cooker, besides its general cleanliness and the absence of fumes.

Fig. 109 shows a small cooker suitable for a small flat or bedsitter. It has two controlled radiants which also serve as the grill. The oven is thermostatically controlled and the door is of the 'drop down' type. The cooker can be operated from a 13 A or 15 A power supply and is available with an optional stand (shown in the illustration) which provides additional storage space. Fig. 100 shows a modern electric cooker with four high-speed radiant rings, a waist-level grill and a family-sized oven. It is fitted with an automatic oven timer, a minitimer and clock.

The desired oven temperature is obtained by setting the control dial at the requisite temperature, which is maintained

Fig. 109 Small electric cooker with thermostatically controlled oven and two radiants which also serve as the grill

(G.E.C. Ltd.)

Fig. 110 Modern family-sized electric cooker with four high-speed radiant rings, an automatic oven timer, minitimer and clock

(G.E.C. Ltd.)

within close limits, as shown by the graph in Fig. 111.

Simple clockwork timers or electric timers are a common feature; they simply cause a bell to ring when the period for which they have been set has expired. Automatic oven timers are designed to control the oven switching for predetermined periods at some time ahead, so that the housewife can set the timer, go out for the day and find the joint ready-cooked on her return. These timers are just a special application of the ordinary time switch, with adjustable cams on discs, driven by an electric synchronous clock, which close and open switch contacts for the oven, the temperature having been pre-set and controlled by the oven temperature regulator,

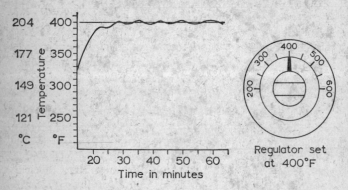

Fig. 111 Oven temperature control by thermostat-type regulator

which is usually of the bimetallic thermal intermittent switching type. This kind of regulator is also used to control hot-plates.

'Plug-in' elements are provided in the majority of cookers and the spaces between the oven walls are lined with heat-insulating material. Such additional features as foot pedals for door opening, high-level grills and rotisseries, indicator lights, heated plate drawers or cupboards, reversible oven doors, height adjustment, castors or rollers for easy removal and an interior glass door to view the contents of the oven without the risk of cooling draughts are provided on some models; a recent innovation is a self-cleaning oven, in which extra heat over a period removes the scale resulting from cooking processes. It is also possible to obtain cookers with the oven at different levels or even as a separate unit, so that a group of hotplates, the oven and the control panel can be fitted into a continuous work-top (see Fig. 128, page 208).

In considering circuit loadings, the maximum load that is likely to be on at one time can be taken to vary from two-thirds for an 8 kW cooker to one-half for a 15 kW cooker. Trends in recent years have been towards higher loadings, so that, whereas in the past a 7 kW family cooker with a 30 ampère supply circuit was considered normal to provide for in a two- or three-bedroom house or flat, it is now necessary to allow for cookers with a total loading up to about 15 kW, and enquiry should be made to ascertain this before wiring. Although cookers vary so much in loading, the cooking requirements of an average family do not vary to the same extent. Even if a cooker that is much too large for normal requirements is chosen and installed, the maximum load taken is not likely to exceed that taken by a smaller cooker by very much, but if and when larger meals are prepared for guests or a party the cooker may be more fully utilised. However, regulation requirements allow for diversity in the use of the various parts of a cooker such that a 30 ampère circuit may still be installed for any cooker with a loading up to 14 kW.

There are two types of boiling plate: the solid plate type, which presents a smooth, black, machined, cast iron surface from which the heat is transferred by conduction to the utensil on top and by radiation from the exposed element underneath to a grill pan below; and the radiant type, with a reflector plate or drip tray on the underside, in which the heat transfer is mainly by radiation but also by conduction where the element is in contact with the utensil. With the former type, utensils having flat, machined bottoms that cannot buckle should be used in order to obtain maximum heat transmission, while with the latter type direct contact is not essential.

The element of the enclosed type is embedded in grooves underneath the cast iron plate and retained in place by a special cement; the cast iron plate must be earthed. The radiant type consists of a nickel-chrome heating spiral enclosed in a steel tube and insulated therefrom by magnesium oxide. The solid plate type is only used in small cookers, such as breakfast cookers; this single hotplate unit can also be used for grilling and it is called a boiler-griller. However, it is heavy, easily corrodes or rusts and is inefficient if good contact cannot be maintained with the utensil (as with light utensils not having heavy machined bottoms). Thus it has generally been superseded for hotplates by the radiant type, which has rapid heating response, is more efficient and has loadings to bring a pint of water or milk to the boil in about 4 minutes. The loading of an 180 mm boiling ring is about 2000 W, while the 150 mm size takes about 1650 W.

The grilling element is often in the form of an open wire spiral in a ceramic former or supported below a boiling plate, and the loading is from 1000 W to 2750 W, depending on the size. Boiling plates and grill elements are separately provided with energy regulator controls, by means of which the heat from the elements can be varied over a wide range from 'simmer' to full heat. This form of regulator operates switch contacts on and off at various intervals, depending on the pressure on a bimetal strip heated by the element current, and is adjusted by the control knob.

Connection of cookers

This should be carried out by a competent installation contractor or by the supply authority's employees if the cooker is on hire purchase. The cooker circuit must be terminated in a control unit with a double-pole main switch, so that the cooker can be completely disconnected from the supply. A three-pin socket may be included in the cooker control unit for an electric kettle. A typical cooker control unit, with neon pilot lamps to indicate when a switch is on, is shown in Fig. 112. In places or installations where the supply capacity or maximum load must be restricted, overloading is prevented by having a control switch of the change-over type, so that either the cooker or the washing machine (alternatively, the immersion heater) may be in circuit, but this type of control unit is seldom necessary in modern installations. The auxiliary socket outlet may be 13 A or 15 A, but in the latter case a fuse unit for the socket is incorporated in the control unit. Cooker control units should comply with

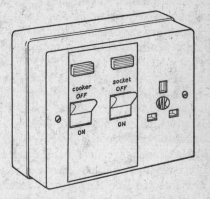

Fig. 112 Cooker control unit with pilot lamps
(M.K. Electric Ltd.)

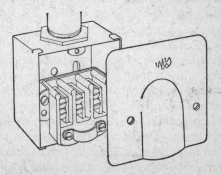

Fig. 113 Flush connector box and surface plate for lead to cooker
(M.K. Electric Ltd.)

B.S. 4177 and are available in surface or flush, all-insulated or metal-case construction. The cooker is connected to a separate circuit in the house installation by a flexible multi-core cable or single–core cables in a flexible tube from the control unit. In flush wiring a separate connector box in the wall below the control unit may be used. This is shown in Fig. 113. Cooker loadings vary according to size from about 2 kW for a small breakfast cooker up to the loadings mentioned earlier. It is essential that the cooker and its control switch are properly earthed, either through the conduit or by means of a separate earthing conductor.

Electric kettles

An electric kettle is an essential adjunct to an electric cooker, as the use of a hotplate for boiling water is less efficient and takes up useful cooking space. Electric kettles have an internal immersion element that gives out all its heat to the water; this is the highly efficient feature of electric water heating that makes it superior to other methods of heating water. The immersion element consists of resistance wire coiled in mineral insulation within a metal tube and terminating in the connector assembly, so shaped that it can be passed into the kettle through the lid opening and the connector can be passed out again through a hole in the back, where a watertight joint is made to the kettle with rubber and fibre joint rings and a screwed outer connector casing. A safety device is usually incorporated to ensure interruption of the supply if the kettle boils dry. This may be in the form of an ejector rod which pushes the removable part of the connector off the kettle, when a fusible joint attached to the kettle element releases a spring, or it may be a thermal cut-out which breaks a contact in the heater circuit, usually by means of a bimetal strip, if overheated. Kettles should always be fitted with three-core flexible cords and three-pin plugs so that they are effectively earthed. The fixed kettle connector must always be in the form of shrouded pins and the loose flexible cord connector should be in the form of sockets, otherwise exposed live pins on this end of the flexible cord would be very dangerous. Some kettles have a thermostatic device that automatically switches the element off when boiling point is reached.

Electric irons

A well-balanced smoothing iron of the correct shape and weight, with a handle so shaped that fatigue is minimised, should be chosen. The most common source of trouble is with the flexible cord, as this is so liable to be twisted and to suffer abrasion in use; therefore, the flexible cord should be

Fig. 114 Easispray Deluxe electric iron with thermostatic control. (A spray, steam or dry iron of modern design. Instant change from steam to dry ironing is achieved by an easily operated slide switch. A pilot light indicates when the selected temperature has been reached.)

(Morphy Richards Ltd.)

frequently examined and replaced when necessary. A good-quality, three-core, unkinkable flexible cord should be used. In use the iron must not be left flat on the table, or it will burn right through the wood, and some fires have been caused by this happening. A 750 W lightweight iron of modern design with thermostatic control is illustrated in Fig. 114.

Steam irons are made so that a small quantity of water is put in an opening with a spring stopper on top of the iron; when hot, steam issues through grooves in the face of the iron near the point.

Electric water heaters

Even though many houses obtain hot water by means of coal fires or slow-combustion boilers, the provision of electric water heaters is an added convenience with high inherent efficiency, automatic control, and freedom from dirt and fumes.

There are five main types of electric water heating equipment:

(a) Storage heaters of the pressure type.
(b) Storage heaters of the displacement (or non-pressure) type.
(c) Immersion heaters (including internal and external circulators).
(d) Instantaneous heaters, or geysers.
(e) Electrode boilers.

Pressure-type storage heaters are usually installed in new houses and when an existing domestic hot-water system is being converted, as they represent the best design of water heater and are the most efficient and the cheapest in the long run.

Immersion heaters are widely used for the conversion of or to supplement existing solid-fuel boiler systems and where low initial cost is of primary importance, but the efficiency of the installation is lower.

Electric instantaneous water heaters give small quantities of hot water immediately. They have a disproportionately high electric loading and are not viewed with favour by supply authorities, due to their heavy intermittent loads. A small unit requires about 3 kW, while for a bath at least 10 kW are needed for a limited flow of water.

The electrode water heater is only applicable where large quantities of hot water are required for central heating or laundry purposes, and it is not used in normal-sized houses. The principle of operation depends on the heating effect of a current passing through the water between a system of electrodes suitably arranged for the supply. This type will not be considered in any greater detail.

Pressure-type storage heater

This type of storage heater is subject to pressure due to the head of water in the cistern, which is 0·1 kg/cm^2 per metre head. The heater consists of an inner container of welded copper, tinned internally where required for drinking and other domestic purposes, with an outer casing of sheet steel, the intervening space being packed with granulated cork or polyurethane foam to provide heat insulation. The smaller sizes are suitable for wall mounting, while the larger capacities are provided with feet for floor mounting. An immersion heater and a thermostat are inserted in the side of the tank at low level. In some water heaters a longer heating element is inserted almost vertically in the top of the tank and reaches to the bottom. In principle, the heated water rises to the top of the tank and increases in depth until it reaches down to the level of the thermostat, which then switches off the supply, but water below the level of the lowest part of the heater element cannot be heated. Hot water is drawn off from the top of the tank and cold water enters at the bottom. This type of water heater is intended for installations with two or more draw-off points with taps.

A modern type of automatic electric storage heater is

shown in Fig. 115. This apparatus has been designed to meet the demand in flats and small kitchens where space is restricted, and the dimensions are such that the UDB-20 can be accommodated in the space under the draining-board. This location is ideal because of the short length of pipe to the sink, where the largest and most frequent demand for hot water occurs. The two sizes, 91 and 136 litres, are 0·84 and 1·22 metres respectively in height, and two separate immersion apparatus plates are provided. The one near the top consists of one 1000 W element controlled by its own thermostat, while the bottom plate has one element of 2000 W controlled by a second thermostat. The top element will provide a constant supply of 32 litres of hot water sufficient

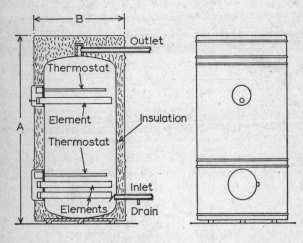

Sectional elevation   Front elevation

Dimensions	A	B
UDB–20/	0.84 m	0.508 m
UDB–30/	1.22 m	0.508 m

Fig. 115 Pressure-type electric storage heater
(Sadia Water Heaters Ltd.)

for washing-up and cleaning purposes, while the bottom element can be switched on when baths are required. All the 1000 W elements are interchangeable and withdrawable when the heater is full; the internal wiring is complete, so that the electrical fitting is confined to the necessary connections to a double-pole switch and the provision of a good earth connection.

The arrangement of this water heater for an independent service is shown in Fig. 116, in which the heater is directly fed

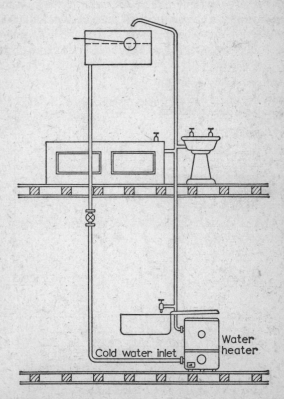

Fig. 116 Pressure water heater and pipework for three outlets

from the balltank. When working in conjuction with an existing hot-water system, the arrangement is shown in Fig. 117. It is fed by hot water from the boiler through the existing hot-water storage tank. This apparatus works at a high all-round efficiency, and it is estimated that, with electricity at $\frac{1}{2}$p per unit, ample hot water can be provided for a family of five for less than 30p per week.

For normal domestic demands in the average medium-sized house, a water heater smaller than 91 litres capacity should not be used unless the requirements are known to be very moderate, as barely two small baths in succession can be provided with this capacity.

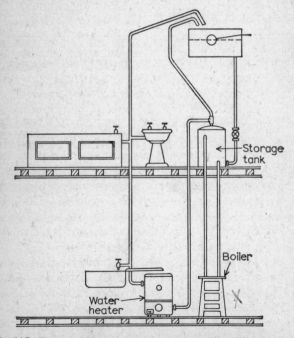

Fig. 117. Pressure water heater in conjunction with an existing hot-water system and storage tank

Table 17 on page 186 gives some idea of the average family consumption of hot water, from which it is possible to work out the storage capacity required, provided that a temperature recovery rate of 2 minutes for a 1 kW heater element per litre of water used is taken into account between draw-off times.

When hot water storage is required at off-peak tariffs, it is important to calculate the amount of hot water required per day, since it is not possible to heat up more water in restricted hours when all the hot water has been used unless means are provided to switch over to the daytime supply at the higher peak-hour tariff rate as an emergency measure. A minimum capacity of about 227 litres is necessary for off-peak storage in the average small household.

Open-outlet types

Self-contained water heaters from 6 to 68 litres capacity to supply one point are operated on the displacement or non-pressure principle and *must be provided with an open outlet* so that pressure cannot build up within the container, which would otherwise explode with the resulting water pressure for which it is not designed. Therefore, a tap must not be fitted to the outlet. When the inlet control tap is turned on, the entering cold water displaces an equivalent volume of hot water through the open outlet spout into the washbasin or sink. Thermostatic control is provided, and the electrical and water connections are quite simple; for the 6, 10 and 13·8 litre sizes (see Fig. 118) direct connection may be made to the water main; these are the usual domestic sizes.

The small instantaneous water heater for washbasins is not uncommon for hand-washing in lavatories; it is usually fitted with a spray nozzle on the swivel outlet, and with a 3 kW loading it delivers about 1·4 litres per minute at 38°C. It is an open-outlet or non-pressure type of water heater and may be connected to the water mains or to a cistern supply.

It is essential, before installing any type of water heater, to

Fig. 118 10 litre, 1 kW open-outlet water heater
(Sadia Water Heaters Ltd.)

see that it is suitable for the hardness and acidity of the local water, and that the water supply authority's regulations are not infringed.

Immersion heaters

There are several different arrangements of heating elements that can be fitted to existing hot-water tanks. They usually consist of a mineral-insulated, copper-sheathed, tubular element and an adjustable or fixed temperature thermostat in a separate tube fitted into a terminal head, with cover and screw boss, ready to insert into a suitable flange mounting, which must be provided at low level in the tank. Such a unit

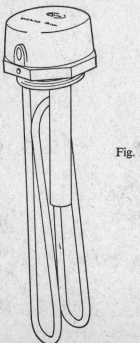

Fig. 119 3 kW immersion heater and thermostat
(Santon Ltd.)

is shown in Fig. 119. This is a non-withdrawable type, because it cannot be taken out unless the tank is empty. Withdrawable immersion heater elements and thermostats are contained in hollow tubes fitted to the head which enable them to be withdrawn with the tank full of water.

Immersion heater elements suffer from furring, just as a kettle does, and they are also corroded by aggressive water, so the element sheaths are made of metals more immune than copper to these effects where the water is cuprosolvent. Plating, monel metal and stainless steel have been used, but the latest metal to be adopted is titanium. The

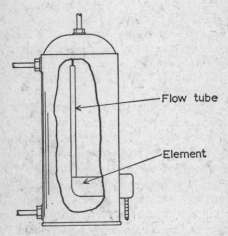

Fig. 120 Side-entry circulator

latter has proved most satisfactory, but, in common with other special metals, it is expensive.

The method of dealing with hard water is to set thermostats at a lower temperature than with soft waters to reduce the furring. With soft water thermostats can be set up to 80°C, but higher temperatures than 60°C are not recommended with very hard water.

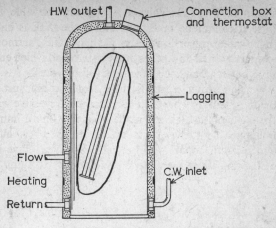

Fig. 121 Top-entry circulator

Electric circulators

This type of heater is fitted vertically, or almost so, in a flow tube to produce convection currents so as to cause rapid circulation of the water. They have the advantage of a much faster recovery rate should all the hot water be used, due to the rapid convection currents, and a quantity of really hot water is quickly restored to the top of the tank. Figs. 120 and 121 illustrate side- and top-entry circulators, but although a number of circulators are in use they are not now made to any extent except by certain Scottish manufacturers.

Conversion of existing hot-water systems

The success of the conversion depends on the condition and layout of the existing pipework, because badly designed hot-water supply pipework can cause serious waste of heat, with unwanted water circulating currents in the pipes, which cannot be afforded with electric immersion heaters. Similarly, when hot-water radiators for space heating are connected to boiler systems, it is both inefficient and uneconomical to allow the electrically heated water to circulate in them. Expert advice should be obtained before an installation is begun. In general, the installation will be satisfactory where:

(a) The tank is efficiently lagged with at least 75 mm of high–grade insulation.

(b) The run of the hot-water pipes is short, not in exposed positions, and so designed as to avoid circulating currents in vertical pipes and reverse circulation in a boiler that is not in use, in periods when no water is drawn off, due to cooling of the pipes in the surrounding air. Bad design of pipework in this respect causes excessive heat loss and consequent consumption of electricity with unlagged pipes.

(c) The hot-water pipes are all connected to the top of the tank or expansion pipe, which should have at least 500 mm horizontal from the tank before changing to a vertical run for the reason given in (b) above.

It should also be recognised that a bare hot-water storage tank in a boiler system provides a 'heat leak' or a means of cooling overheated water when the fire burns fiercely and no water is being drawn off. This waste cannot be afforded, nor is it necessary with electricity. So, to compensate for the tank lagging that must be provided, it is a good idea to provide an alternative heat leak for the boiler only (separately connected to the boiler) in the form of a radiator or heated towel rail, which becomes an additional amenity.

Settings for thermostats

Although less water can be stored at higher than at lower temperatures for mixing with cold water to obtain the same quantity of water at the required usable temperature, it is not advisable to set thermostats too high, since scale deposit or 'furring' of the immersion element becomes serious at high temperatures. In addition, the standing heat loss from the water cylinder may become excessive.

With soft water, and up to about 12° hardness, the thermostat may be set to interrupt the supply at a water temperature of 80°C; with medium hardness, at 70°C; and with hardness over 20°, the thermostat should not be set over 60°C.

Water-heating data

With water heating we are able to calculate the power required with a greater degree of accuracy than with space heating. We know that one unit of electricity (1 kWh) = 3·6 MJ and that 1 litre of water (1 kg) requires 4180 J to raise the temperature 1°C. Therefore, the consumption of electricity will be:

$$\text{Number of units (kWh)} = \frac{\text{Litres} \times \text{Temperature rise °C}}{860 \times \text{Efficiency}}.$$

Allowing for losses, the efficiency varies from about 92 % with good lagging to less than 80 % for unlagged tanks. From the above equation the size of heater required to heat a given quantity of water in a given time may be ascertained. Assuming, for easy calculation, an efficiency of 87 %, we obtain 750 in the denominator; then

$$\text{Size of heater in kW} = \frac{\text{Litres} \times \text{Temp. rise °C}}{\text{Time in hours} \times 750}$$

$$\text{and litres} = \frac{\text{kW} \times \text{Time in hours} \times 750}{\text{Temp. rise °C}}.$$

A room temperature of 60°F is approximately 15°C.

Cold-water inlet temperatures may vary from 5° to 21°C, depending on the locality and the time of year, but 5°C is used for water-heating calculations.

Use	Litres required	Required temperature	Litres required	
		°C	60°C	71°C
Hot bath	114	44	74	61
Hand wash	4·5	44	3·1	2·4
Dishwashing	4·5/meal	60	4·5	3·7
House cleaning	9/day	60	9·9	7·5
Laundry	45/week	60	45·5	37

Table 16 Hot water used in average household

Note: The above figures assume a yearly average cold water temperature of 12°C.

Various temperatures (*Fahrenheit and Celsius*)

	°F	°C
Boiling water temperature	212	100
Scalding temperature	158	70
Washing-up dishes, temperature	140	60
Bath, average temperature	104	40

Table 17

Approximate capacity of tanks
Cylindrical tank = (Diameter)^2 × Height × 785·4 litres.
Rectangular tank = Length × Breadth × Height × 1000 litres.
All the above tank dimensions are in metres.

As a rough rule, 1 unit of electricity will boil 8 litres of water starting from cold.

Example 16. A cylindrical tank is 300 mm diameter and 650 mm high. How many litres will it hold, and what size of immersion heater should be fitted if all the water is to be heated from 5°C to 70°C in 2 hours?

Cylindrical tank = $0·3^2 × 0·65 × 785·4 = 46$ litres

Size of heater $= \dfrac{46 × (70-5)}{2 × 750} = 2$ kW.

8 Layout of the Installation

The first thing to do is to mark a plan of the house with symbols of the electrical equipment required and then to prepare a corresponding schedule. If a plan of the house is available, the details should be laid out on this, or a sketch plan should be drawn approximately to scale.

Symbols for plans

British Standard Graphical Symbols are given in British Standard Specification No. 3939, and an extract of Section 27, covering symbols commonly used in interior wiring installations, is shown in Fig. 122 (page 188). These symbols are used on the plans of the houses illustrated.

It will often be found that architects and installation engineers employ their own symbols, especially on small-scale plans where the standard symbols cannot be clearly marked, but such deviations should be minimal.

Installation materials

When the installation has been planned, the next consideration is the standard of installation to adopt. The use of steel conduit for wiring in houses as the highest standard has long been regarded as best practice; even where cheapness was important, steel conduit was used with close-joint and split-type or grip-joint conduit fittings. With the development of plastics, however, plastic conduit has become much more suitable for house installations, and steel is now regarded as more appropriate for factory installations, public and other

No.	Description	Symbol	No.	Description	Symbol
27.2	Lighting points or Lamps		27.5	Control and Distribution	
27.2.1	Lighting point or lamp: Add lamp details, e.g.: 3×40 W		27.5.1	Main control of intake point:	
27.2.2.	Lamp or lighting point wall mounted:		27.5.2	Distribution board or point: Show circuits controlled by qualifying symbol or reference: e.g. heating: lighting: ventilating:	
27.2.4	Lighting point with built-in switch:				
27.2.5	Lamp on variable voltage supply:				
27.2.8	Single fluorescent lamp:		27.5.3	Main or sub-main switch:	
27.2.9	Group of fluorescent lamps: or:	3×40 W	27.5.4	Contactor:	
			27.5.5	Integrating meter:	
27.3	Switches and Switch Outlets		27.5.8	Transformer:	
27.3.1	Single-pole one-way switch (several indicated by a number):		27.5.9	Consumer's earthing terminal:	E
27.3.4	Cord-operated single-pole switch:		27.6	Fixed Apparatus and Equipment	
27.3.5	Two-way switch:		27.6.1	Electrical appliance: (designate type):	
27.3.6	Intermediate switch:		27.6.2	Fan:	
27.3.7	Time switch:		27.6.3	Heater: (specify type):	
27.3.8	Switch with pilot lamp:		27.6.4	Motor:	M
27.3.10	Dimmer switch:		27.6.6	Thermostat:	
27.3.11	Push button:		27.6.11	Bell:	
27.3.12	Luminous push button:		27.6.12	Indicator panel (N = number of ways):	N
27.4	Socket outlets:		27.6.13	Clock:	
27.4.1	Socket outlet:				
27.4.2	Switched socket outlet:				
27.4.4	Socket outlet with pilot lamp:				
27.4.5	Multiple socket outlet: e.g. for three plugs:	3			
27.7	Telecommunication Apparatus including Radio and Television				
27.7.1	Telephone call point:		27.7.8	Radio or television receiver: (state service):	
27.7.4	Socket outlet for telecommunication: e.g. television: radio: sound:	TV R S	27.7.9	Amplifier:	
27.7.6	Aerial:		27.7.10	Microphone:	
27.7.7	Earth:		27.7.11	Loudspeaker:	

Fig. 122 A selection of British Standard symbols

large buildings, and places where the strength and rigidity of steel is required. Conduits have the greatest advantages in long life, ease of wiring alterations or replacements and cable protection; but, for lowest cost and the simplest form of wiring, the sheathed cable system is very largely adopted nowadays in buildings with wood-joist floors.

The advent of plastics has also affected switchgear and other wiring accessories. Switchgear for most house installations is obtainable with an all-insulated plastic enclosure, and this type of equipment is most appropriate for sheathed wiring or plastic conduit systems, although metal-clad switchgear and fuseboards are often used instead, especially where larger sizes and capacities are not available in the all-insulated types.

Plastics have also largely superseded rubber for cable insulation, and standard p.v.c.-insulated cables are made for up to 600 V to earth and 1000 V between conductors to B.S. 6004, and with butyl rubber insulation to B.S. 6007; p.v.c. and vulcanised rubber are used for most of the domestic flexible cords.

Whereas with steel conduits these are used as the earth-continuity conductor, with plastic conduits it is necessary to include a separate earth-continuity conductor in all the conduits, together with the circuit wiring; with plastic-sheathed wiring, this must also have a separate earth-continuity conductor within the sheath.

Layout considerations

The lighting in a new house should be considered in relation to the furniture arrangements in the various rooms. The switches should not be behind doors or in inconvenient positions just to save a few feet of wiring. The position of lighting points and switches should be carefully selected before the wiring is begun so as to avoid later alterations with sunk work; such alterations, involving replastering, often cause unsightly discoloration of the wallpaper. The occupier should be consulted if possible, but experience and common sense should be used in offering advice.

The service cable will be arranged by the supply undertaking, and their fuses and meters should be adjacent to the point of entry but should not take up valuable cupboard or pantry space. With some modern houses the garage is often convenient for this, or a special cupboard or inset recess can be provided which will also house the main distribution boards for lighting, power and other purposes. It is a growing practice to arrange such a position with a small window or covered opening to the outside so that the meter-reader can see the meter without having to enter the house.

We will now look at the installation for two different houses. The first one is a detached house in which no solid fuel or gas is used and electricity serves for all purposes. The second installation is for a small semi-detached house, and in this case the electrical installation is on a more modest scale, as solid fuel is used and the house is piped for gas. In both cases the loadings are arbitrary, as the user will no doubt choose lamp sizes to suit himself or the fittings to be installed, and these details are not usually known at the design stage.

Electrical installation of a detached house
The load is obtained from the schedule of lighting and sockets, which also shows other information and lighting data for the various rooms. The sizes of socket outlets have been confined to 13 A, and these would be wired on ring circuits. The use of twin 13 A socket outlets for all positions in the kitchen, main rooms and bedrooms would be very advantageous and would cost very little extra. The electric clock points, shaver sockets and the bell transformer take only a small current, so they have not been included. The positions of switches and various outlets are indicated in the plan (Fig. 123, page 193) by British Standard symbols, but the

Schedule of lighting and sockets

	Size m	Area m²	Lamps	Watts/ room	13 A sockets
Dining room	5·0 × 3·5	17·5	3 × 60 W(T)	180	5
Music room	3·7 × 4·1	15·2	1 × 150 W(T)	150	5
Study	2·6 × 3·0	7·8	1 × 100 W(T)	100	3
Hall	2·4 × 3·5	8·4	1 × 60 W(T)	60	1
Lobby, lavatory, W.C.			3 × 40 W(T)	120	
Kitchen & cup- boards	3·8 × 2·4	9·1	2 × 40 W(Fl) 2 × 40 W(T)	180	4
Wash-house	1·8 × 1·7	3·1	1 × 60 W(T)	60	1
Tool shed	1·8 × 1·2	2·2	1 × 40 W(T)	40	
Porches			2 × 40 W(T)	80	
Garage	2·6 × 4·5	11·7	2 × 60 W(T)	120	1
Ground floor		75·0		1090	20
Bedroom 1	5·0 × 3·5	17·5	1 − 60 W(T) 3 − 40 W(T)	180	4
Bedroom 2	3·8 × 4·2	16·0	1 − 60 W(T) 2 − 40 W(T)	140	4
Bedroom 3	3·8 × 2·5	9·5	1 − 60 W(T) 2 − 40 W(T)	140	3
Sewing room	2·4 × 3·0	7·2	1 − 40 W(Fl)	50	2
Bathroom	2·5 × 2·1	5·2	1 − 60 W(T)	60	
W.C.			1 − 40 W(T)	40	
Top of stairs landing			2 − 40 W(T)	80	1
Airing cupboard			1 − 40 W(T)	40	1
First floor		55·4		730	15
Total		130·4		1820	35

Details of room sizes, lighting and socket outlets

conduit runs are not shown. It is a good idea to arrange the larger rooms so that they are not dependent on one sub-circuit; and, if the house load necessitates more than a single phase of the supply to be brought in with load balancing on two or three phases, it is important to avoid outlets on different phases being on the same floor, and certainly not in the same room, as line voltage (415 V) would exist between conductors possibly within reach and therefore dangerous.

The supply authority's fuses and meter are shown in the garage at A, mounted at a reasonable height on the wall and adjacent to the incoming service cable. The consumer's main switches and fuses can also be at A. An alternative point of entry is shown to the left of the front door, with the distribution fuseboards at B at the back of the hall. The final choice of positions depends on local requirements. In either of these alternatives, it would be possible to arrange the meters behind a glass panel or small window in the wall to enable the meter-reader to inspect without having to enter the house.

Wiring sizes and loading

The total general lighting load is 1820 W, which with a 240 V supply gives a current of 7·6 A, though all this will never be on at one time. Although 2 A socket outlets could be used for lighting purposes instead of some of the 13 A sockets shown and be wired to the lighting circuits, the saving would be small and the usefulness of 13 A sockets for all purposes would be lost.

Considering the ground floor, the maximum lighting load is 1090 W, which requires 4·5 A at 240 V. As there are nine-teen points, two circuits should be run from the distribution boards with 1 mm² cables.

Reference to Table 6 in Chapter 4 (page 79) will show that voltage drop calculations may be ignored, as the maximum length of run of the circuits is so short. There are twenty 13 A sockets, and these may be connected to a single 30 A ring

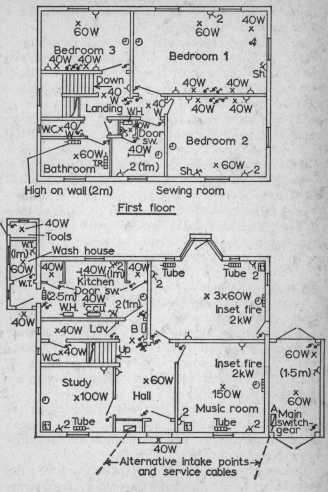

First floor

Ground floor

Fig. 123 Plans of a detached house

circuit, which may supply an unlimited number of 13 A
sockets within an area of not more than 100 m² in domestic
premises with 2·5 mm² rubber- or p.v.c.-insulated cable, or
1·5 mm² mineral-insulated cable.

On the first floor the maximum lamp load is 730 W, which
requires 3 A at 240 V. There are sixteen points, and again two
circuits are desirable to avoid the whole floor being in darkness if a fuse 'blows'. Regulations limit the number of lighting
points on a circuit to a maximum of 15 A load, with each
lamp rated at 100 W. Thus the smallest p.v.c. cable of 1 mm²
section in a group of three with a capacity of 9 A could feed up
to twenty-one lamps. There are fifteen 13 A sockets, which
may be connected to a single 30 A ring circuit.

As this will be an all-electric house, consultation with the
occupier and the electricity supply authority will probably
result in the adoption of an all-in tariff, and there will be no
need to keep lighting, heating and power points in separate
installations for separate metering. With a single-phase, two-
wire service, therefore, four lighting circuits, two ring
circuits for socket outlets and separate circuits for an electric
cooker and water heater, one distribution fuseboard with six
15 A ways (one spare and four fused 5 A) and three 30 A ways
will be adequate for the whole house and, if expense will
permit, a miniature circuit-breaker distribution board will
provide a higher standard of installation.

Details of lighting, switching and socket outlets
The dining room is shown with a central fitting for general
lighting, containing three 60 W lamps controlled by two
switches for using one, two or three lights. The switch is
mounted near the door, at the same side as the handle,
1·3 m above the floor. A common height should be maintained throughout the house. 13 A socket outlets are provided on each wall between the doors and the fireplace to
avoid flexible cords crossing them, and there is a built-in
electric fire which can be connected to the ring circuit

through a fused spur unit and a separate double-pole switch if the switch is not incorporated in the electric fire itself. The other fixed heaters can be connected by socket outlets, or in the same way as the fixed fire, but with pilot lamps where necessary. The sockets should be fixed towards the top of the skirting board to give room for the flexible cord to turn out from the bottom of the plug. The recommended minimum height above floor for sockets is 150 mm, but it has always been felt that 450 mm above floor level is ideal and avoids reaching to floor level to insert plugs or operate the switches of switched sockets. For use with bench or worktop equipment, the sockets should be 150 mm above bench level. A permanent fixed clock point is provided above the fireplace and is connected to a lighting circuit; this obviates trailing leads to the synchronous clock and is also indicated in other rooms.

The music room has a 150 W central fitting with two-way switching adjacent to each door, and the previous remarks about socket outlets also apply to this room. The use of extra standard-lamps is intended to boost the lighting locally as required in both these rooms.

The study is provided with a 100 W central fitting, but if the walls are dark or there are many bookshelves a greater wattage would be needed; it is assumed, however, that a desk or table lamp would be used here to supplement the general light. The positions of the various outlets depend on the wishes of the occupier and the distribution of the furniture. No bracket lighting has been indicated, but decorative wall bracket lights are extremely popular in living and drawing rooms, and suitably located points for these should not be overlooked. Sockets for the vacuum cleaner are provided in the ground-floor hall and first-floor landing.

The kitchen has two 40 W fluorescent lamps in enclosed fittings, which, it is suggested, should be positioned to give good illumination to the cooker and sink on one side and to the work table between the larder and a cupboard which gives access to the serving hatch on the other. In each of these cupboards a 40 W lamp is mounted in a ceiling fitting, controlled by door switches which go 'on' when the doors are opened. A similar arrangement is fitted to the airing cupboard on the first floor. The main lighting is controlled by two two-way switches at each door.

The wash-house has an enclosed fitting, the switch being watertight; the porch light is an enclosed bracket fitting with the switch in the kitchen.

The hall and first-floor landing are provided with 60 W and 40 W lamps respectively, both with two-way and inter-mediate switching.

The garage is provided with a general light (which might be better placed at the end where the car bonnet will be) con-trolled by two two-way switches, a light over a bench and a socket for a portable electric tool or handlamp.

The bedrooms are provided with a 60 W lamp over the dressing table, and 40 W lamps over the beds and in front of a wardrobe; alternatively, wall brackets or portable bed-table lamps can be provided and connected to twin or separate socket outlets by the beds. The switches are shown near the door, on the same side as the handle, and two-way switching is provided from the bed and door positions. Ceiling switches can be used for the wardrobe light, and bed switch and wall fittings over bedheads instead of ceiling points if desired.

The bathroom is provided with a 60 W enclosed ceiling fitting controlled by a cord-operated ceiling switch. If the bathroom is to be provided with a washbasin and mirror, an all-insulated wall lighting fitting is desirable over the mirror, and possibly a shaver supply unit as well.

The illumination of the sewing room should be generous, hence a 40 W fluorescent lamp is used with socket outlets for a portable lamp and the sewing machine, if an electric model is used.

The other lighting arrangements do not call for any special comments, except that the wash-house and porch fittings should be of the watertight pattern, and that the sizes of lamps given can be increased with advantage if higher standards are desired.

Schedule of heating and other outlets

It is convenient to consider other fixed heating and power outlets separately, and a schedule of these is given on page 198; the various outlets are indicated on the plan (Fig. 123, page 193). The 13 A socket outlets listed in the lighting schedule will provide for all the portable apparatus likely to be used, including portable fires and kitchen equipment, but there are fixed fires, tubular heaters and other non-portable apparatus which should be listed. With the exception of the cooker, these outlets are connected through switched fused spur units to the nearest ring circuit for 13 A sockets, as shown in Fig. 58, page 88, using a suitable size of cable for the apparatus to be connected but ensuring that the spur fuse is not too large to protect the cable and, in any case, not larger than 13 A rating.

In the dining room and the music room 2 kW built-in fires are fitted, while a socket outlet in the study can be used for a 2 kW portable fire, convector or fan heater if desired. In the three main ground-floor rooms, tubular or skirting heaters are fitted under the windows, each consisting of a bank of two tubes 1·2 m long taking 480 W, or equivalent, which on a 240 V supply requires 2 A, so 1 mm² cable is large enough. Thermostats are mounted in the corners of the rooms on the outer walls so as to switch these heaters off when the room temperature exceeds 15°C. When the other electric heating is not in use, these tubular heaters keep the rooms aired in cold weather. A similar arrangement of heaters could be adopted for the bedrooms, but our scheme assumes the use of portable heaters in these rooms. A portable convector heater can be used in the hall. The most popular method of warming the bathroom is to use a so-called 'infra-red' heater, as illustrated in Fig. 103, page 149, fitted

300 mm below the ceiling and efficiently earthed. An electric towel rail, 750 mm long and loaded at 75 W, is provided. These points can be connected to the sockets' ring circuit through switched spur units. All the exposed metalwork of electrical equipment, and the metalwork of other services also, must be electrically bonded together in the bathroom.

Schedule of heating and other outlets

Room	Heaters	Other apparatus
Dining room	1 2 kW fire 2 tubular, 960 W	
Music room	1 2 kW fire 1 tubular, 480 W	
Study	1 tubular, 480 W	
Kitchen	Wall heater, 750 W	Refrigerator cooker, 12 kW water heater, 1500 W
Wash-house		Washing machine with heater 2500W
Bathroom	Towel rail, 75W 1 wall heater, 750 W	
Linen cupboard		Water heater, 1500 W

The kitchen contains the cooker, which is fed from a separate fuse in the consumer unit and for which the load is ascertained by taking the sum of the first 10 A of the total cooker loading, 30% of the remainder and 5 A for the socket outlet in the control unit. This results in a capacity of less than 30 A for the control unit and circuit cable, with the exception of the largest domestic cookers, for which 45 A rating is more suitable. This circuit is wired back to a separate fuseway in the consumer unit or, if necessary, to a switch fuse unit at the main switchboard. A pair of 6 mm² p.v.c. cables will carry 31 A, but 10 mm² cable may be necessary for the larger cookers. The 90-litre water heater, rated at 1500 W, takes 6·25 A and is connected to a switched spur unit, with a pilot lamp fed from the nearest socket ring circuit, or to a separate fuseway in the consumer unit using 1 mm² cable, thus maintaining greater capacity in the ring circuit.

If extra heating is required in the kitchen for cold mornings, a portable heater can be used, or a high-level heater could be fixed on the wall—as for the bathroom—and a suitable outlet point provided. The refrigerator is connected to a 13 A fused plug fitted with a 3 A fuse.

The separate water heaters for kitchen and bathroom would be more economical in saving heat losses in long runs of pipework if a single water heater were installed. The kitchen water heater also supplies the adjacent lavatory basin, and the bathroom water heater in the airing cupboard could also supply basins in the bedrooms if desired, but the design of an electric water-heating system is a special subject and is referred to in Chapter 7.

If a water-heating cylinder is not installed in the linen cupboard, a small tubular heater would be required here.

The wash-house is provided with a watertight 13 A socket outlet mounted well above the floor for the wash-boiler.

Estimation of current

All this load will not be on at once, and diversity of use can be allowed for in determining the capacity of the main cables and switchgear. This is done as follows, using the allowances for diversity given in the I.E.E. Wiring Regulations:

		Ampères
Lighting:	60% of 1820 W = 1214 W:	5
Cooker:	say, 12 kW; first 10 A:	10
	30% of remainder:	12
	If socket-outlet in control unit:	5
Water heaters: 100% of first two 3000 W:		12·5
Socket outlets and stationary appliances:		
	100% of largest fuse rating of individual circuits:	30
	40% sum of remaining fuse ratings:	12
	Total:	86·5 A

Main supply and load balance

Having estimated the total electrical load in the house, the supply authority must be approached and asked whether this load can be connected to a single-phase, two-wire supply. If so, this figure can be rounded off to 100 A for the capacity of the main switch fuse or circuit-breaker, and

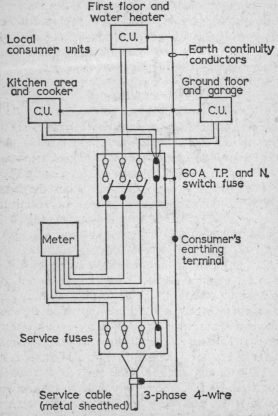

Fig. 124   Main supply and distribution switchgear for three-phase distribution in house

the main cables must therefore not be less than 35 mm^2 in section for p.v.c.-insulated cables. In the event of the supply authority requiring the load to be balanced on two or three phases of the supply, the whole distribution system would have to be designed differently and divided into two or three independent sections with equal loads, as far as practicable, but so arranged that the wiring to all outlets in the same room are on one and the same phase. When possible this rule should be applied to a whole floor to ensure maximum safety and the avoidance of shock at medium voltage (415 V). Fig. 124 shows how such an installation is arranged for balancing on a three-phase, four-wire supply.

Installation of small semi-detached house

The two floor plans of this house are shown in Fig. 125. The outlets are indicated by standard symbols, which show the minimum installation requirements. An electric cooker and an immersion heater for the hot-water tank are included, but coal fires are available in four rooms for heating. There is a coke boiler in the kitchen, which also warms the bedroom above by the hot-water pipes to the tank in the linen cupboard. Electrical appliances are so commonly used in even the smallest households that it is false economy and unrealistic to limit unduly the number of socket outlets, and therefore the number shown here should not be reduced. One ring circuit is utilised with universal 13 A socket outlets for both floors, and where the ring is not conveniently looped into some of the socket outlets on a different floor they can be connected as spurs from the ring. Twin socket outlets provide numerous additional points without extra wiring to avoid the use of loose adaptors. These sockets should be fixed not less than 150 mm, and preferably 450 mm, above floor level, except in the kitchen, where they should be 150 mm above bench level. In most other respects, what has been said previously regarding the example of the large house installation will apply to the small house installation.

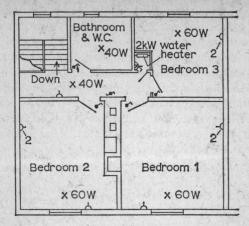

First floor

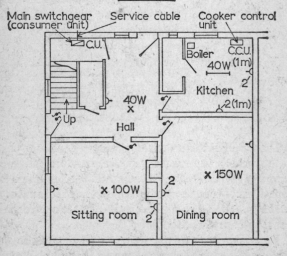

Ground floor

Fig. 125 Layout of small semi-detached house

Two sub-circuits for lighting are run: one up to the loft, dropping down to the bedrooms, and the other going to the space below the first floor to drop down to the ground-floor fittings.

The incoming supply runs along the back of the houses, the service cable entering beneath the stairs.

The schedule of points is given below, from which it will be seen that the connected lighting load is 600 W, allowing for 10 W loss in the 40 W fluorescent lamp. The supply is 240 V a.c., which gives a maximum current of 2·5 A, but this is not demanded in practice.

Total current

The total load is calculated as for the previous example as follows:

		Ampères
Lighting:	66% of 600 W = 400 W:	1·7
Cooker:	say, 8 kW: first 10 A:	10
	30% of remainder:	7
	If socket-outlet in control unit:	5
Immersion heater: 100% of 2 kW, say:		8·3
Socket outlets: largest fuse rating:		30
	40% of remaining fuse ratings:	10
	Total:	72 A

The supply authority might consider a greater diversity to be applicable and might install a standard single-phase service of 60 A capacity, but the main switchgear and cables capacity should be based on the above figures. Thus the main switch fuse should be 100 A capacity and the main cables not less than 25 mm² section.

The cooker takes the heaviest unit load and requires 4 mm² cable, but it would be wise to install 6 mm² cable in case a larger cooker is required in the future. The separate circuit for the immersion heater should be run in 1·0 or 1·5 mm² cable. The wiring can be in plastic conduit for a first-class job, but plastic-sheathed, multicore cable would probably be used.

The ring circuit would consist of 2·5 mm² with spurs, while the lighting cables would be 1 mm². In a small house like this with short runs, there is no need to worry about voltage drop calculations.

Lighting schedule

Room	Size metres	Area m²	Lamps	13 A sockets
Dining room	5·0 × 3.2	16	150 (T)	3
Sitting room	3·8 × 2·8	10·6	100 (T)	3
Kitchen	3·0 × 2·0	6	40 (Fl)	4
Hall	—	—	40 (T)	1
	Ground floor		340 W	11
Bedroom 1	3·8 × 3·0	11·4	60 (T)	3
Bedroom 2	3·8 × 2·8	10·6	60 (T)	3
Bedroom 3	3·0 × 2·6	7·8	60 (T)	2
Bathroom and W.C.	—	—	40 (T)	—
Stairs and top landing	—	—	40 (T)	1
	First floor		260 W	9
		Total	600 W	20

With an 'all-in' tariff the consumer unit would have three 15 A and two 30 A fuseways—the 30 A fuseways for the ring circuit and the cooker—one 15 A fuseway for the immersion heater and two fused at 5 A for the lighting circuits. This does not permit of any extensions, which, if provided for, need a spare fuseway.

With some authorities the cooker and immersion heater (or washboiler) are connected to a changeover switch to limit the load, so that the immersion heater cannot be on at the same time as the cooker, and vice versa. The immersion heater would have thermostatic control and a separate switch in the airing cupboard.

Variations in domestic installation wiring

A development of the spur from ring circuits for other 13 A sockets is to feed the lighting circuits from spur boxes fused at 3 A, taking the lighting wiring up through the switches to the light fittings on walls or ceilings, as shown in Fig. 126. This reduces the length and cost of lighting circuits, and the size of the consumer unit, but it is not advisable for the

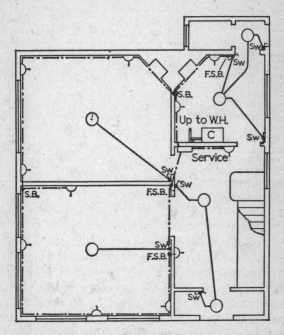

S.B.• Spur box   F.S.B.• Fused spur box
———— Wiring run above ceiling
—·—·— Wiring run at low level
════════ Wiring to cooker and water heater

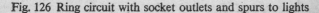

Fig. 126 Ring circuit with socket outlets and spurs to lights

larger installation, where it is best to keep the lighting completely separate from the socket-outlet circuits. Water heaters and other fixed appliances may also be connected to spurs from the ring circuit.

In some installations with the distribution point in a central position, it may well be more convenient and economical to use radial circuits for the 13 A socket outlets instead of the ring circuit.

A method of wiring more suited to large housing schemes than to individual or small numbers of dwellings is the harness system, whereby the wiring for a house or flat is designed and constructed in the factory, so that the wiring 'harness' is placed in position, and the outlets and consumer unit fixed and connected with the minimum of labour on site. This arrangement usually involves a central junction box placed in the floor or roof space, with radial circuits of sheathed wiring or cables in plastic small-bore tubes from this box to all the equipment and accessories of the installation.

The kitchen

The kitchen is the housewife's workshop and many different labour-saving devices will be required, together with an adequate supply of hot water, an electric cooker, a refrigerator, an electric washer, a drying cabinet, spin drier, dish-washer, waste disposal unit, mixer, percolator, an electric clock with timing device, even a carver and a radio.

These are only some of the items now available, with the possibility of more in the future. The layout of the kitchen should be such that it is planned to give the most complete service with ample and convenient outlets for connection. This cannot be achieved with only one or two socket-outlet points fitted at positions that are assumed to be suitable. An ideal arrangement would be a pair every metre around the kitchen, and this idea is already available, because neat wall trunking fitted with 13 A sockets at any required spacing is used in science laboratories at schools and research establishments, and could easily be adapted to domestic kitchens.

The provision of complete kitchen units which can be assembled on site is also available, and these could have similar facilities built into them as well as local lighting.

The layout of kitchen furniture and apparatus must be co-ordinated under the following headings:

1. Relative disposition of equipment for operations.
2. Basic requirement of apparatus, e.g. water, electricity.
3. Objective visibility under natural and artificial lighting.
4. Safety and convenience with minimum fatigue in use.

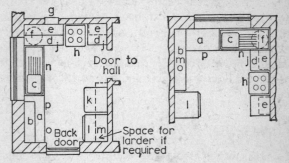

(a) L-shaped layout   (b) U-shaped layout

Fig. 127 Efficient planning of essential kitchen equipment

Key:
a Preparation counter
b Ventilated food store over
c Sink
d Service and dishing-up counters
e China and glass store over
f Water heater under
g Service hatch
h Cooker
j Pans and utensils under
k Washing machine
l Refrigerator
m Grocery store
n Garbage can under sink or waste disposal unit
o Vegetable store under
p Space for drier

(A dishwasher might stand over *f* or under draining board)

Fig. 128 Modern kitchen with built-in cooking units

Classic layouts following the basic preparation-cooker-sink disposition of equipment with L-shaped and U-shaped planning are shown in Fig. 127, but lighting, socket outlets, electric clock, etc., are not shown for clearness. Fig. 128 shows what a modern fully equipped kitchen looks like where there is room for a central unit and cooking units can be built into the furniture units instead of being separate, free-standing equipment. The use of new materials and installation methods, together with regulations that consider convenience, adequacy, good technique and safety, enables much of the drudgery of housework to be removed at a cost that is not prohibitive.

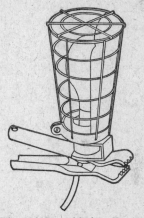

Fig. 129 Portable handlamp

Garage equipment

All garage socket outlets should be of iron-clad type mounted above bench level. Earthing is very important where used for small portable tools such as drills and grinders. All-insulated or double-insulated portable tools are becoming increasingly available; they are quite safe and obviate the necessity of earthing, but care is necessary to keep them in good condition and to avoid damage to the outer insulating shell, which could make unearthed and exposed metal parts actually dangerous in the event of a fault in the equipment. Portable handlamps should be of the all-insulated type, in which it is impossible to touch any metal part of the lamp cap. Transformer equipment is available to provide low–voltage lighting for handlamps which makes portable apparatus absolutely safe to use, and the transformer steps the voltage down from 240 to 12 V. A portable handlamp with a strong wire guard is illustrated in Fig. 129, but a safer pattern is of moulded insulating material. The gripping attachment is useful as it leaves both hands free for working. Flimsy flexible should not be used; either t.r.s. or a good-quality workshop flex should be fitted and examined periodically to see that it is in good condition.

9 Transformers, Electric Motors and Motorised Appliances

Transformers are used on alternating current supplies to obtain changes of voltage and current with the same amount of power. A transformer is a static piece of apparatus consisting of a laminated iron core on which is wound either two separate windings, i.e. a double-wound transformer, or a single coil with a tapping point brought out, i.e. an auto-transformer, as illustrated in Fig. 131, page 213.

A double-wound transformer is represented in Fig. 130,

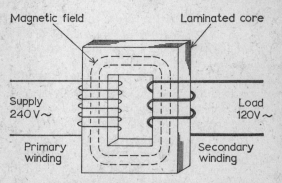

Fig. 130 Diagram of a double-wound transformer

with an alternating current supply of 240 V applied to the left-hand coil, called the primary winding. An alternating magnetic field is set up in the iron core which induces an alternating voltage in the right-hand coil, called the secondary winding. With half as many secondary turns as on the primary winding, 120 V will be obtained across the secondary terminals. If this winding is connected to a load, the current will be approximately twice the value of the primary current. Transformers have very small losses and the voltage is proportional to the number of turns in each winding, while the current is inversely proportional to the turns ratio.

Transformers with iron cores are used for low-frequency applications, i.e. on mains supplies and with low audio-frequencies in radio and television receivers, but at high radio-frequencies the iron core is omitted. Small transformers for domestic bell and other extra-low-voltage circuits are mounted in either an iron or a bakelite case and should be properly earthed. They are only suitable for alternating currents, and if connected to a direct current supply of the same voltage they will take a higher current, get very hot and burn out, if the circuit fuse does not do so first.

Autotransformers

Autotransformers, which consist of a single winding on an iron core with a tapping at the required voltage, must not be used for extra-low-voltage bell and other circuits, as the lower voltage winding is not entirely separate and insulated from the main supply. The lower voltage winding of an auto-transformer should be adjacent to the neutral connection, and the common terminal of both windings should be connected to the neutral pole of the supply. Fig. 131 shows an autotransformer *incorrectly connected* and the dangerous condition it creates. The live conductor is shown connected to the common terminal for both sections, and it will be seen that, even though there is only 6 V, for example, across the bell circuit, this circuit is at least 234 V above earth. This voltage is dangerous both to the extra-low-voltage bell wiring and to any user of the equipment. N terminal should be the common terminal for both sides, so that the extra-low-voltage connections are at the earthed neutral end of the winding.

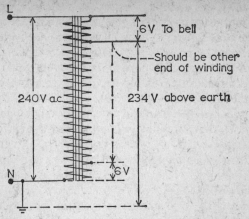

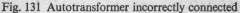

Fig. 131 Autotransformer incorrectly connected

Electric motors

For domestic applications electric motors are of small size and are usually fractional horsepower motors.

These small machines can be switched directly on to the supply but must be suitably wound for the type of supply and mains voltage. Direct-current motors will not operate on alternating current, and vice versa, in general.

There is one type, however, called the universal motor, that will run on either d.c. or a.c. of similar voltage. Such motors do not exceed about 200 W and the majority of domestic motors are below this output.

Under the metric system, motor power output is measured in kilowatts or watts, and the equivalent rating of the many old $\frac{1}{4}$ h.p. rated motors still in service is approximately 186 W.

Principle of the electric motor

An electric motor is used to convert electrical energy into mechanical energy and thus perform work at a certain rate. When a conductor is carrying an electric current, it exhibits

magnetic properties. If another magnet system is adjacent to
this conductor, there will be a mutual reaction and the wire
will be forced to move in a certain direction. By a suitable
arrangement of an iron electromagnet, which is generally
stationary, and a rotating part, called the armature, which
carries a suitable winding, mechanical power is obtained.

The outer stationary frame of an a.c. motor is called the
stator, and the rotating part (termed the armature in a d.c.
motor) is called the rotor. The main windings are on the
stator, but the rotor may have a winding like a d.c. armature
connected to a commutator or slip rings, or a number of bar
conductors with a short-circuiting ring at each end, in which
case it is called a squirrel-cage rotor.

Protection of small motors

The greater the power output of a motor, the more current it
will take from the supply. If more work is demanded, the
motor will take a greater current; if an overload is excessive,
too much heat is generated and the speed of the motor may
fall. To protect the motor, fuses, a thermal device or a small
circuit-breaker may be employed which opens the circuit
when the current becomes excessive or an overload persists
for too long. A small, double-pole, miniature circuit-breaker
which provides automatic overload protection to small
motors is illustrated in Fig. 132. The fixed current setting
gives protection that cannot be modified by unauthorised
persons. The moulded enclosure affords protection from
injury, and the 'free handle' mechanism prevents the circuit-
breaker from being closed on a fault.

Types of d.c. motors

D.c. motors are mainly used for extra-low-voltage battery
operation in this country, where all mains supplies are a.c.,
with the exception of a few very remote parts, but their
type-names and characteristics are given as follows. The
three main types of d.c. motors are (i) shunt, (ii) series and

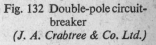

Fig. 132 Double-pole circuit-
breaker
(J. A. Crabtree & Co. Ltd.)

(iii) compound, depending on the arrangement of the field magnet windings. The shunt motor runs at practically constant speed, and its high-resistance field winding is connected in parallel with the armature winding.

The series motor has low-resistance field coils connected in series with the armature. The speed varies with the load, increasing as the latter falls, but it will exert a much greater turning moment, or torque, than the shunt motor at starting and low speeds.

The compound motor is a combination of these two types and is used when the load demands a higher starting torque than the shunt motor without the wide speed variation of a series motor. One application is driving the compressor of a refrigerator.

The connections of these motors are given in Fig. 133,

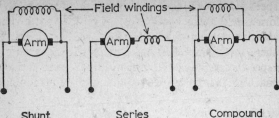

Fig. 133 Connections of d.c. motors

while Fig. 134 shows characteristic curves of speed and torque against current.

With d.c. motors the armature winding is connected at intervals to the commutator copper segments, which are insulated from one another and bear carbon brushes which form sliding contacts.

Wear of the brushes, dirty commutator and poor connections are the usual sources of trouble in small motors.

Universal series-type motors

These motors can be used on either d.c. or a.c. supplies of the same voltage. The entire magnetic system is laminated and

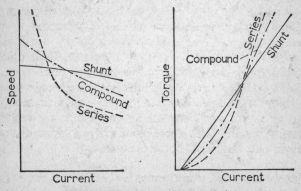

Fig. 134 D.c. motor characteristics

the speed varies with load, as is usual with series motors. The performance is not as good on a.c. as on d.c., but it is quite suitable for vacuum cleaners, small fans, hair driers, floor polishers, mixing machines and other applications where the high speed is acceptable. The armature is the same as that of a d.c. machine, but the brushes require more frequent renewal.

Types of a.c. motors

The a.c. series motor is similar to the universal motor and has already been described.

The basic principle of the static transformer, i.e. mutual induction between the windings, is the same principle that operates the induction motor, which can operate on three-, two- or single-phase supplies. Even though small three-phase induction motors have been developed, the single-phase induction motor is used for domestic applications, as only a single-phase supply is generally available. The single-phase motor is not inherently self-starting but will continue to run once it is started. In the split-phase type an auxiliary winding artificially makes the motor 'two-phase' during the starting period, while in the shaded-pole motor a copper shading ring around half the pole gives the necessary starting torque by causing time displacement of part of the magnetic circuit. This method can only be used for very light starting torque.

Split-phase motors are suitable for applications where heavy starting torques and overloads do not occur, e.g. electric washing machines fitted with a clutch. The artificial second-phase winding of a split-phase motor must be energised in a different period of time from the main single-phase winding, and to do this a capacitor is connected to the second- or starting-phase winding. Shaded-pole motors are used for small slow-speed fans. These motors have a 'shunt characteristic', i.e. the speed is nearly constant.

The repulsion-start, induction-run motor. This has one winding, connected to the supply on the stator, with an armature

like a d.c. motor, but the brushes are connected together. It
starts as a repulsion motor with a 'series characteristic', and
when it has run up to some 75% of full speed the armature
winding is short-circuited by means of a centrifugal device
and contacts on the end of the rotor, and the brushes are
lifted. It then runs as an induction motor.

The capacitor motor. The capacitor-start, induction-run
motor has two windings on the stator, one of which is
connected to a condenser or capacitor, which in effect makes
it a two-phase motor for starting. The auxiliary winding and
capacitor are cut out at a predetermined speed by a centrifugal switch and the motor runs as an induction motor. In the
true capacitor motor, capacitors are switched into a different

Fig. 135 Capacitor motor with
         resilient mounting

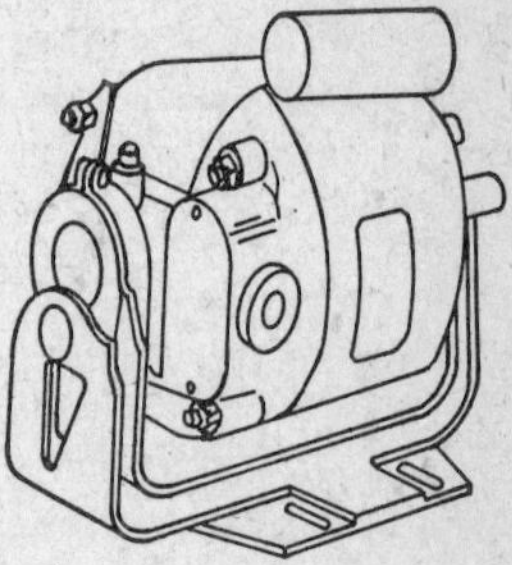

grouping and retained in circuit continuously after running
up to speed, so that the motor runs under load as an artificial
two-phase motor thereafter with enhanced performance.
The rotor is of the robust squirrel-cage type; there is no
electrical connection between the stator and rotor in either
type. Although primarily developed for refrigerators and oil
burners, these motors are extremely well suited to other
duties requiring operation for long periods where maximum
efficiency is essential. A capacitor-start motor is illustrated
in Fig. 135. This motor has a resilient mounting. The robust
rotor construction is evident from Fig. 136.

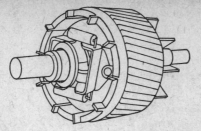

Fig. 136 Rotor of capacitor motor showing centrifugal switch-gear on shaft

This is a most satisfactory type of single-phase, low-wattage motor, even though it costs slightly more than the ordinary split-phase motor. The relatively high efficiency and power factor is indicated in Fig. 137, together with the torque. Capacitor motors are inherently quiet, free from radio interference and economical in running cost.

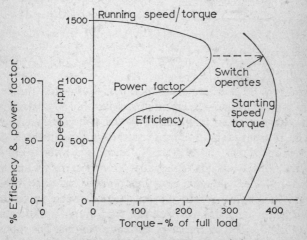

Fig. 137 Capacitor motor characteristics

Synchronous motors. Synchronous motors run at a constant speed, exactly in step with the supply frequency, whereas the motors previously described lose about 4–5 % of the synchronous speed (which is called the 'slip') and are termed asynchronous motors. The domestic application is chiefly for electric clocks that operate on the 50 Hz (c/s) supply and are called synchronous clocks. One type requires the spindle to be given a twist to start. Another type makes use of the shaded-pole principle and is self-starting; the special shape of the rotor ensures that it runs at synchronous speed. The synchronous speed depends on the number of pairs of magnet poles on the motor and is independent of the voltage. Series motors and various a.c. commutator motors can run at speeds above and below synchronism. Table 18 below gives the synchronous speed for various numbers of poles on a 50 Hz (c/s) a.c. supply.

No of poles	2	4	6	8	10	12	16	20	24	30	40
Rev/min	3000	1500	1000	750	600	500	375	300	250	200	150

Table 18 Synchronous motor speeds on 50 Hz (c/s)

Vacuum cleaners

The working parts of a vacuum cleaner consist of a high-speed motor with a shaft extension at one end, to which is attached a fan. The motor is of the universal type already described and may run as high as 10 000 revolutions per minute. From the table above it will be seen that the maximum speed for a synchronous or induction motor with one pair of poles is 3000 rev/min on 50 Hz (c/s).

There are two types of vacuum cleaner, according to the arrangement of the dust bag. In one, the dust bag is attached to the handle and the dust passes through the fan; in the other, the dust bag is of a special fabric inside a canister, which allows the suction of the fan to draw the dust into the bag, the air being discharged over or through the motor or used at that end for 'blowing' instead of 'suction'. In the former type, it is common to have also a rotating brush driven by a rubber belt from the motor shaft. The most frequent troubles are blockage of the vacuum hose in the latter type and damage or kinking of the trailing flexible lead, which should be frequently inspected for abrasion or damage to the insulation. The motor brushes require periodical renewal; and, if the motor will not start, a test should be made to see if there is a break in the series circuit. This might be due to a broken wire or to bad contact at the switch, brushes or terminals.

Laundering appliances

Electric washing machines are constructed of welded sheet steel with a stove-enamel finish. In general, they usually have a rotary agitator or gyrator in a tank, a pump and hose connections—one to the hot water tap and the other to waste at the sink. A heater for the water can be provided as an alternative to a hot-water supply. In older types a vertical power drive was provided for the attachment of a wringer or ironer, but the wringer has now been superseded by the spin drier and the ironer is an independent unit with its own motor drive. The pump empties the tank through a hose in a few minutes. Washers such as that shown in Fig. 138 are made to take up to 5 kg of clothes and larger machines can also be obtained. The motor is a 200 W, single-phase, a.c. induction motor (or shunt-type motor if d.c.). Automatic safety devices are also usually incorporated for protection against mal-operation. To obtain trouble-free service over many years and to maintain the washing machine in good condition it should be regularly inspected by an authorised dealer at least once a year, besides carrying out the simple maintenance prescribed on the instruction sheet.

Modern forms of automatic washing machines may have permanent fixed plumbing connections for water supply and

Fig. 138 Fully automatic washing machine   *(Hoover Ltd.)*

automatic controls to time the washing, rinsing and drying operations, and to regulate water temperatures to suit various fabrics. Some machines use the washing tub to run at high speed for spin drying.

One of the most popular types of machine is the combined washer and spin drier mounted side by side in the same cabinet with hose connections to the sink and hot tap.

Spin driers

Spin driers have internal perforated containers which run at a high speed of about 600 rev/min and throw the water out of the wet contents by centrifugal force in a few minutes, leaving the fabrics almost dry enough for ironing. Capacities up to 2·7 kg of clothes are available. Tumbler driers are also made, in which the clothing is tumbled over at much lower speeds with heated air flowing through the fabrics, and this gets clothing even drier than the spin drier—ready to iron or wear. They usually have a glass front-opening door for inspection. Automatic control may be provided to regulate the period of drying, and all models have switch-interlocked front or top lids to stop the machine if the lid is opened during operation.

Rotary ironer

The most popular type of rotary ironer is provided with a 500 mm or 650 mm long padded roller, rotated by a small motor, and an electrically heated pressure shoe of stainless steel. The smaller size is more in demand as it is much lighter for handling and can stand on a bench or table. The ironing shoe has a loading of 1000 to 1500 W according to size, usually with an energy regulator to control the iron temperature. Electric hand irons and other domestic heating equipment are described in Chapter 7.

The domestic refrigerator

A refrigerator must have efficient thermal insulation; the door must be tight fitting so that there is no air leakage; and the internal air circulation must allow the air to flow easily over the cooling surfaces. Easy cleaning is essential, and the parts containing the refrigerant must be of robust construction so that there is no possibility of the escape of gases.

The electrical refrigerator can be operated by either the

mechanical or compression system, or the non-mechanical or absorption system. In the former a motor-driven compressor is used, while in the latter an electric heater is used to circulate the refrigerant round the system.

Absorption refrigerators are made for operation by electricity, gas or oil. The inherent efficiency of the absorption system is much lower than that of the compression system; but they are practically noiseless, whereas the motor and compressor are not. Modern domestic refrigerators of the compressor type are so quiet, however, that this is not usually a problem and the majority sold are of this kind.

Principle of operation

When a liquid is vaporised, heat is required (i) to raise its temperature to boiling point, called the sensible heat, and (ii) to convert the liquid to gas at the same temperature, called the latent heat.

If the liquid is kept in a vessel at a high pressure, the boiling point is increased by the pressure; if the high-pressure liquid is allowed to pass into a vessel at a much lower pressure or a vacuum, it immediately turns into gas, but at the same time heat is abstracted from its surroundings.

Refrigerants used

The same principle applies to both types of refrigerators, but in the compression type freon is used, while the absorption type uses ammonia as the refrigerant, with hydrogen as the inert gas in the evaporator. In either method the evaporator is maintained some 11°C below the temperature required in the refrigerator; thus heat passes from the air and food in the cabinet to the evaporator coils, causing any liquid refrigerant in the tubes to change to gas.

Operation of compression refrigeration

A simple diagram is given in Fig. 140, in which the evaporator coil is inside the cabinet while the condenser coils are outside,

Fig. 139 Modern domestic refrigerator

(Electrolux Ltd.)

either below or at the back, adjacent to the motor-driven compressor. The evaporator coil is around the ice-making compartment trays at the top, and the cold air falls to the lower spaces and is replaced by warmer air, from which the heat is abstracted. A single-cylinder, single-acting compressor is shown; the suction valve opens slightly before the piston reaches the bottom of its stroke and gas enters the cylinder. When the piston begins to rise, both valves close and the gas is compressed; at a certain pressure, the delivery valve opens and the high-pressure gas passes to the condenser. A slight loss of heat liquefies the gas, the cooling process being assisted by cool air from the fan blowing across the cooling fins. The final cooling is carried out at the bottom of the condenser, which contains high-pressure liquid. The compres-

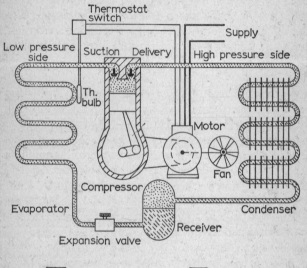

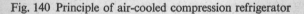
░░ Gas   ▨ Liquid

Fig. 140 Principle of air-cooled compression refrigerator

sor cylinder is kept free of liquid and only deals with the gas. The gas volume is very much greater than an equal weight of liquid, so a high-pressure receiver is included in the circuit to contain the liquid, above which is space for some high-pressure gas. The receiver is connected to an expansion valve, which allows some liquid to pass to the evaporator, with loss of pressure and absorption of heat, thus changing into gas. The cycle of operation is then repeated.

Temperature control

A spirit-filled, sensitive, phial-operated thermostat controls the interior temperature within an adjustable range. The thermostat bulb is against the evaporator, and a capillary tube goes to the flexible copper bellows which operate the thermostat switch to start and stop the motor. The refrigerating unit's capacity is more than ample for producing the necessary refrigeration, so the motor is automatically on intermittent duty. With the absorption type, the refrigeration is more nearly continuous and there is not the same reserve of capacity.

The motor

Frequent starting and stopping, and ample starting torque for the compressor, make the capacitor motor most suitable, as it gives good starting torque, is silent in operation and free from radio interference. For d.c. supplies a compound motor is used. Domestic refrigerators need a motor output of from 120 W to 200 W depending on the cubic capacity of the refrigerator, which ranges from about 0·03 to 0·34 m^3 in the usual domestic sizes. Small models use about 350 units, and the 0·17 m^3 size about 500 units per annum. The electric refrigerator is an example of precision engineering, and in the unlikely event of failure expert attention should be obtained.

Modern refrigerators use more advanced types of compressors than the one illustrated, but the principle of refrigeration is the same.

The freezing cabinet

This modern development of the refrigerator is gaining popularity in the home for storing large quantities of perishable food for long periods at lower temperatures than are obtained in the common domestic refrigerator. It operates in the same way, however, and incorporates a larger capacity machine with more power for producing sub-zero temperatures.

10 Primary Cell and Accumulator Batteries

A battery is a group of cells joined up to give greater power; the cells are generally connected in series, so that the individual voltage of each cell is added to the next one, as explained in Chapter 2. Cells of different sizes should not be connected together, since their internal resistances differ. Primary cells transform chemical energy into electrical energy. The Leclanché cell is generally used for bells and other small-current intermittent work; the chemical constituents are the same in both the wet and the dry form. The wet Leclanché cell consists of a glass jar containing a solution of sal-ammoniac (ammonium chloride), in which stands a porous pot with a central carbon plate, which is surrounded by the depolariser. The terminal on the carbon is the positive, while a zinc rod in the electrolyte is the negative element. The electrolyte and the zinc rod are expendable and are easily renewed. Wet cells are best kept in a cool place to minimise evaporation; dry cells, on the other hand, are better in a slightly warm situation. The dry cell has the same chemical constituents as the wet cell, but the chemical paste, which takes the place of the sal-ammoniac solution, cannot be renewed; also, the outer zinc case becomes perforated, due to the chemical action, and cannot be replaced like the zinc rod of the wet-type cell. An inert cell is a form of dry cell that can be stored for long periods without deterioration. When required, a little water is introduced into a vent hole, thus moistening the chemical paste. The e.m.f. of these cells is about 1·5 V; it falls if excessive current is taken from them, depending on their size, but they are quite satisfactory for intermittent work.

For continuous duty and heavier currents, secondary cells or accumulators are used. An accumulator has plates of certain chemical composition which is changed by the action of the charging current, which passes from the positive plate to the negative plate through the electrolyte. When charging is complete, the reverse action is possible, the accumulator can supply a current and the plates revert to their original condition. These processes can be repeated without renewing the constituents of the cell, which must be done with a primary battery.

There are several types of accumulator, but the most common types used for domestic purposes are:

(a) The lead-acid cell, with lead plates treated with certain oxides in a solution (electrolyte) of dilute sulphuric acid.
(b) The nickel-alkali cell, which has nickel-iron or nickel-cadmium plates with an electrolyte of concentrated potassium hydroxide.

New types of cell are being developed for electric vehicles on the one hand and for spacecraft on the other. The former include zinc/air, sodium/sulphur and lithium/chloride as electrode-active material combinations, but they are not generally available yet. New types of small light cells for light current purposes are also barely out of the research laboratories. The latter types are of limited application, mostly in research and circuitry in aeronautical and space exploration equipment, and in special miniature electronic instruments.

Cells for small batteries are usually of the sealed type, having totally enclosed tops with vent plugs or stoppers for access to the electrolyte. This reduces evaporation considerably and prevents spillage.

Lead-acid cell

The lead-acid cell is made up of pasted plates, the positive being red lead and the negative litharge mixed with sulphuric acid. An initial forming process changes these plates to chocolate-coloured lead peroxide and spongy lead of a grey metallic colour. The brightness of these colours is one of the indications that such a battery is well charged and in good

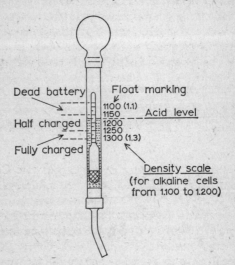

Fig. 141 Hydrometer for lead-acid accumulator

condition. When the terminals are connected to an external load, the electrical potential existing between them forces a current to flow round the circuit, chemical changes occur, the plates go dull and the density of the electrolyte decreases, due to the formation of water. The density of the dilute sulphuric acid used varies with the type of cell but is generally around 1·21, and a measurement of the density with a hydrometer, illustrated in Fig. 141, is a good guide to the amount of charge left in the battery. Some accumulators have a floating device that shows the state of charge, while large storage batteries installed in power stations have hydrometers that float permanently in the acid.

The electromotive force (e.m.f.) of a freshly charged battery cell is about 2·2 V when it is not supplying current; but when discharging the potential difference (p.d.) quickly falls to about 2 V, where it remains for some time, depending on the rate of discharge, and then gradually falls to such a low value that is useless and cannot support the current required. A lead-acid cell should never be discharged below 1·8 V, otherwise insoluble lead sulphate forms on the plates. This is evident from white patches on the plates, which in time will ruin the battery.

Care of lead-acid accumulator

The following points, if observed, will prolong the life of a battery:

1. Never leave the battery in a discharged state. Charge the battery if the density falls below the minimum stated on the label (usually 1·150).
2. A voltmeter reading across the terminals, on open circuit, is deceptive; it may be 2 V, but when discharging it may fall below 1·8 V.
3. Keep to the recommended charge and discharge rates specified by the maker.
4. The level of the electrolyte must be kept above the top of the plates by adding distilled water to make up for evaporation. Concentrated sulphuric acid should not be added. In mixing the electrolyte, the acid should always be added to the water; the other way round is disastrous.
5. All connections should be kept clean and well smeared with vaseline.
6. Do not short-circuit the battery with a piece of wire;

you may burn your fingers, and, in any case, a heavy
short-circuit current will loosen the paste and may
buckle the plates.

Nickel-alkaline cell

The positive plate in each case is nickelic hydroxide held in a
grid of nickel, while the negative plate is a special mixture of
iron oxide or cadmium carried in nickelled-steel plates. The
electrolyte is potassium hydroxide (caustic potash), which is
a corrosive liquid with a specific gravity about 1·18. The
electrolyte does not change in density, so only very occasional make-up with a little distilled water is necessary.

The average voltage of an alkaline cell is 1·2 V, so more
cells are required to make up a given battery voltage than lead
cells with a p.d. of 2 V. The advantages of this cell are its
light weight, ability to withstand mechanical abuse and
freedom from 'sulphating', and that it can be overcharged,
overdischarged and left uncharged for long periods without
damage. The disadvantages are higher initial cost and the
need for more cells to make up a given voltage.

Capacity of a battery

This is given on the label in ampère-hours at the 10-hour
rate, e.g. a 40 Ah battery means that it will supply 4 ampères
for 10 hours, if in good condition. With higher discharge
rates, the capacity is reduced. With a lead cell, if the current
in this example is increased 50 % to 6A, it will only last for 6
hours before the p.d. falls to 1·8 V, and with currents below
4 A the ampère-hours obtained are somewhat above the
rated capacity. The number of ampère-hours on discharge
compared with that required to recharge the cell (ampère-
hour efficiency) is about 90 % for a lead cell and 75-80 % for
an alkaline cell, though the latter has less difference in
capacity with different discharge currents. Comparative
characteristic curves are given in Fig. 142.

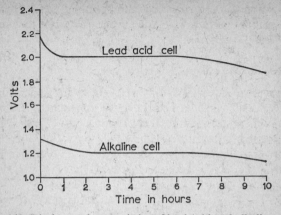

Fig. 142 Discharge characteristics of lead-acid and alkaline cells

Methods of charging accumulators

Accumulators can be charged from d.c. supplies by inserting a resistance in series to 'drop' the difference of voltage, with the required charging current. But this method is wasteful, as the following example will show.

Example 17. A 12 V car battery is charged at 15 V 5 A from a 110 V d.c. supply. What resistance must be connected in series? What will be the loss in the resistance and the cost of charging for 24 hours, assuming the mean charging current is 4 A and electricity costs 1p per unit?

Volts drop in resistance = 110−15 = 95 V.

Value of resistance $= \dfrac{95}{5} = 19\ \Omega$, to carry 5 A.

Loss in resistance (I^2R) = $5^2 \times 19$ = 475 W
(compared with 15×5 = 75 W input to battery).

Cost per 24 hours $= \dfrac{110 \times 4 \times 24}{1000} \times 1p = 10 \cdot 6p$.

Another method is to use a motor generator set, consisting of either an a.c. or a d.c. motor driving a low-voltage generator suited to the batteries to be charged. This method is carried out either by the constant-current method, in which batteries with similar charging currents are grouped together in series and the generator voltage is adjusted to keep the current constant, or by the constant-potential method, in which batteries of similar voltages are connected in parallel across bus-bars, the charging current being large to begin with and decreasing until charging is complete. These methods are only suitable for many batteries where skilled attention is available.

When the lead-acid cell is fully charged, it 'gases' and naked lights or sparking connections or contacts must not be allowed near the battery, as the hydrogen evolved forms an explosive mixture with air.

Houses in remote places without mains supplies may have private generating plants with petrol-engine-driven d.c. generators and large-capacity accumulator batteries which can supply lighting etc. in the house for long periods between charging periods. Such installations must have proper switchboards with instruments and automatic charging cut-outs.

Large establishments where the public is admitted and other places where maintenance of supply is essential if the main supply fails are usually provided with large-capacity batteries for emergency stand-by supplies. These batteries usually supply a minimum degree of lighting for a few hours and are generally 'trickle' charged by charging equipment from the mains to keep them in fully charged condition ready for an emergency. These larger battery installations require skilled maintenance and attention to keep them in good working order; they are not treated in further detail as they are considered to be outside the scope of this book.

The car battery in service is charged by a special type of generator driven by the engine, and the circuit includes a 'cut-out', which completes the charging circuit when the voltage of the dynamo is high enough to provide a charging current and switches off when low engine speed causes the dynamo voltage to fall too low, so preventing the battery from discharging through the dynamo.

For the small accumulator and car battery, static types of charger are most suitable. These are usually of the valve-rectifier type for the larger heavy-current chargers used in garages and works where battery-driven trucks or trolleys are employed, and of the dry-plate rectifier type for small light-current and 'trickle' chargers commonly used by car owners for keeping their car batteries charged when out of use for several days.

The notes on simple circuits and connection of battery cells in Chapter 2 are equally applicable here, so they are not repeated; however, they may usefully be reviewed by the reader, as recognition of polarity is important with accumulators. The positive terminal of the charging circuit must be connected to the positive terminal of the battery and, likewise, negative to negative. Alternating current cannot be used directly for charging batteries, because an accumulator requires a continuous direct current to charge it whereas alternating current reverses its direction in each alternation; so it has to be 'rectified'—i.e. only the positive half of the wave is fed to the battery, or by more complete rectification the negative half is brought over to the positive side, as indicated in Fig. 143. This unidirectional current is quite satisfactory as the charging current and, although not strictly continuous, has an effective value equal to the mean value of the current half-waves. The equipment consists of a transformer to reduce the a.c. mains voltage to the charging voltage, and the secondary circuit includes a rectifier pack in the circuit to the output terminals. A fuse is fitted in the primary circuit and a thermal cut-out is included in the output circuit to protect the rectifier and windings in the event of short-circuit. The small car battery chargers have a screw-

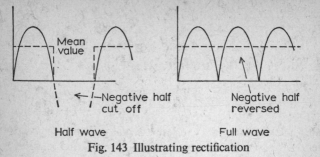

Half wave   Full wave
Fig. 143 Illustrating rectification

plug adjustment for the mains voltage side for a range of a.c. voltages from 200 to 250. On the output side, similar adjustment is provided for 2 V, 6 V and 12 V batteries. The charger is supplied with a length of cable and spring clips to clamp onto the battery terminals. It is important to ensure correct polarity of connections, positive to positive and negative to negative, by colouring or suitable marking of the leads.

Emergency lighting units

Developments in electronics and solid-state circuitry have enabled a new application of small secondary cells to be made in the independent, self-contained emergency lighting fittings that are now available. Although intended for the safety of the public in stores, schools, hospitals, etc., there is no reason why they should not be used in a private house if thought desirable.

A small diffusing bowl encloses two small lamps, a sealed three-cell, nickel-cadmium, alkaline battery, a trickle charger and mains failure relay in a remarkably small space. With a power consumption of only 5 W from the mains, it automatically provides a small amount of light, enough to find one's way about, for a period of 1 to 3 hours if the mains supply breaks down, and practically no maintenance attention is required. Larger units are also available and some contain miniature fluorescent lamps.

11 Radio and Communications

Radio phenomena is very different from ordinary electrical phenomena such as that explained in Chapter 2 and requires an entirely different approach, although extensions of the same theory apply. We have learnt that an unchanging direct current produces a steady magnetic field, but that no effects are induced in other conductors in the vicinity until the current is switched off and drops to zero; this sudden change does, however, produce an inductive effect both in the circuit itself and in adjacent conductors, particularly if the circuit is an inductive one, i.e. a coil with a magnetic field. Alternating current, which is continuously varying, therefore always involves inductive effects, whether in a single straight conductor or in a coil. Inductance depends on the number of linkages with a magnetic field; it is represented by L and is measured in henries (symbol H). It results in a self-induced voltage, diametrically opposed to the applied voltage and lagging a quarter cycle (90°) behind the current cycle (as shown in Fig. 144 (a)) and a reactance, X_L, which is expressed numerically by $2\pi fL$ ohms (f = frequency, Hz).

Similar considerations apply to the stressing of an insulating medium between conducting surfaces, i.e. the capacity effect. No current flows in a condenser or a capacitor with an unvarying potential applied to it, but with an alternating potential the insulating medium is stressed alternately in both directions according to the frequency and a current flow is effected. The capacity is measured in farads (symbol F) and is represented by C. It relates the charge of electricity to the voltage it produces. The numerical considerations are

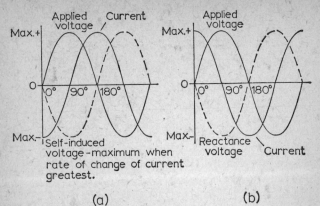

(a)   (b)

Fig. 144 Current displacement with inductance and capacity

    (a) Inductance only
    (b) Capacitance only

exactly as for inductance effects, but the self-induced voltage is 90° in front of or leads the current cycle (as shown in Fig. 144 (b)), so its effect is directly opposite to inductance and its reactance, Xc, is expressed by $\dfrac{1}{2\pi fC}$ ohms.

With such very small radio-frequency currents, the resistance of aerial and feeder conductors is of very little account and is neglected, the main factors being inductance and capacity with their resonating effects—analogous to the tuning fork and vibrating strings in musical instruments; the relation between these two factors is expressed by the impedence value, $Z = \sqrt{\dfrac{L}{C}}$ ohms, as a measure of resonance.

It will be seen that both inductive and capacity effects are a function of the alternating frequency. At low values, such as mains supply frequency, the effects are limited to the local circuits, but as the frequency is increased these effects have much greater electromagnetic effects on the medium known

Wavelength λ metres

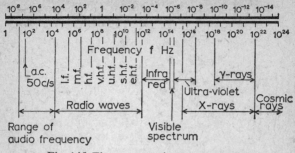

Fig. 145 Electromagnetic wave spectrum

as the ether, which permeates all space and matter but is most responsive in space and the atmosphere. The ether has a high degree of elasticity and response to high frequencies, especially those above 1000 Hz; in free space the oscillations or waves will travel unimpeded for millions of miles, but on earth they are attenuated or absorbed in varying degrees by the atmosphere and other objects in their vicinity. The electromagnetic wave spectrum is shown in Fig. 145, and with our ready knowledge of the behaviour of light rays in the visible part of the spectrum we can begin to understand to some extent how the other kinds of waves at different frequencies operate. Since all radiation waves travel at the speed of light, which is 300 000 000 metres per second, the wavelength and frequency can be related by dividing the velocity by the frequency, i.e. the wavelength (symbol λ, the Greek letter *lambda*) $= \dfrac{300\,000\,000}{\text{Hz}}$, or, since we are dealing with high values of frequency, $\dfrac{300\,000}{\text{kHz}}$ or $\dfrac{300}{\text{MHz}}$.

Having arrived at some understanding of what radio waves are, we can proceed to discuss the method of propagating and receiving them for broadcasting purposes. The aerial is the instrument used for stressing the ether and making it oscillate at the desired frequency to carry the signal it is required to broadcast, and for receiving the signals. Transmitting and receiving aerials are exactly the same in principle and operation, and any aerial can be used for both purposes.

Radio aerials

By imposing a suitable voltage on an aerial system, radio waves will be generated in the ether with a wavelength depending on the frequency of the input and the dimensions of the aerial. The aerial is designed to be in 'tune' or to resonate with the applied frequency by making the length of the aerial conductor correspond to the length of the half-wave so as to be able to carry each full half-wave of current from zero to maximum and then to zero again over its length in alternate directions with the radio wave. The inductance of the aerial distributed over its length and its capacitance between the two halves are calculated to resonate or be in tune with the required frequency and wavelength. Thus maximum signal strength or gain and selectivity (of wavelength) is said to be obtained, and this is augmented by suitably directing the aerial towards the transmitting station.

The common basic aerial design is the half-wave dipole, which is formed of two metal rods or tubes in line, each a quarter wavelength long and connected at the inner ends to the two conductors of the downlead or feeder cable. The dipole aerial is fixed vertically for vertically polarised transmissions and horizontally, broadside on to the transmitting station direction, for horizontally polarised transmissions. The Band II radio signals are horizontally polarised, as are some broadcasts in other bands; others are vertical. Table 19 gives the frequency and wavelengths of broadcast bands used in the U.K.

The list of broadcasting stations within each group is too long to include here, but aerials are made for each group or frequency band to give reasonably good results for any frequency within the group, and for best results a separate aerial should be provided for each group required. Groups of four channels are allocated to each of the sixty-eight stations in the u.h.f. television bands for different programmes; twenty-six stations now transmit I.B.A., 625 line colour/ monochrome pictures in the U.K.; and many others will follow later.

Aerial groups	Channels	Frequency bands	Approximate wavelength, m	Term
Sound radio	—	kHz 150–285	2000–1185	Long wave (l.f.)
,, ,,	—	535–1600	560–188	Medium wave (m.f.)
,, ,,	—	MHz 2–30	150–10	Short wave (h.f.)
Band I television	1 to 5	45–67	6·7–4·5	v.h.f.
Band II radio (programmes 2,3,4)	—	87–100	3·4–3	,, (f.m.)
Band III television	6 to 13	180–215	1·65–1·4	,
Band IV television	21 to 34	470–582	0·64–0·52	u.h.f.
Band V	39 to 68	614–854	0·49–0·35	,,

Table 19 Frequency and wavelength of broadcast bands in U.K.

Vision and sound signals for the same broadcast are transmitted at slightly different frequencies, being only about 3 MHz apart; each station has about 8 MHz bandwidth per channel, and this has been worked out to enable reasonable selectivity and freedom from interference to be obtained from the standard aerials for each band. The frequency and wavelength for each transmitting station refers to the carrier wave on which the signal is imposed by variations of amplitude or frequency, hence the terms amplitude modulation (a.m.) and frequency modulation (f.m.). The power output of small broadcasting stations may vary from 5 to 50kW and provides reasonably good reception within a radius of about 40 to 80 km, but the main stations have powers from 100 to 1000 kW and cover much larger areas, up to 160 km radius or more, according to the height and contours of the area.

Good reception of radio and television transmissions depends primarily on locality and distance from, as well as the output strength of, the transmitting station. Hills, valleys and large structures have a marked effect on the strength of signals received, and site tests of signal strength are often the only way of determining this before a decision can be made on the type of aerial installation to use in areas remote from stations. Tall structures obstruct or reflect signals according to their position in relation to the direct line between station and receiving aerial; this effect may involve the adjustment of aerial direction and is often the cause of 'ghost' or multiple images on the television screen, even in good reception areas.

In some urban areas television and radio relay services are available. In this system a central receiving station has an efficient and well-sited aerial array and amplifying equipment, and redistributes the broadcasts over land lines to subscribers in the area—much in the same way as the Post Office telephone service. The wires are taken into each subscriber's house and a programmes' selector is provided. Receivers may be hired or purchased. Individual aerials are therefore not necessary with radio relay services.

No one will dispute the ugliness of numerous aerials on the roofs of rows of houses, and in estates where this is not permitted, or in blocks of flats where it may be impracticable, central aerial systems and amplifying equipment are installed and cables distribute the signals to the tenants' houses, where their own receivers can be used in the same way as an independent aerial. Subscription is, of course, necessary for upkeep of the equipment.

Most houses, however, have independent aerials, and these are designed, firstly, to suit the waveband of the required transmission and, secondly, to suit the location and signal strength available. The shorter waves are more like light waves and rapidly weaken as they go out; they pass through the ionised layers of the upper atmosphere— whereas the long waves are reflected—and are reflected by hills and tall structures. So, the higher the aerial, the better the reception will be, although in areas within a few miles of the transmitting station good reception is often obtained with a suitable aerial mounted in the loft or roof space, rather than outside on a chimney stack, which makes for a better appearance of the house with cheaper fixings and less maintenance.

Aerial design

The basis of aerial design is to provide a conductor of almost half-wave length, as this resonates to a maximum with the carrier waves and is said to be in 'tune' with the wavelength. In practice, the length is a compromise since it must receive the different sound and vision wavelengths from the same station simultaneously, as well as signals from other stations in the same waveband. This is assisted by making the aerial conductor as large as possible in section to widen its response to waves within the required bandwidth, and for this purpose it should be about 15 mm diameter at least and is usually made from light alloy tube. Aerial makers generally add the channel number or a letter to their catalogue numbers for convenience to designate each kind of aerial; these are as follows:

A for channels 21 to 34 colour code red
B „   „   39 to 51 „   „ yellow
C „   „   50 to 66 „   „ green
D „   „   49 to 68 „   „ blue
E „   „   39 to 68 „   „ brown

The basic dipole aerial can be fitted with what are known as parasitic elements to improve the gain and the directional properties of the aerial. These elements have no direct electrical connection to the aerial of any importance but serve to reflect or direct signal strength to the aerial proper, according to whether they are situated behind or in front of it relative to the signal direction. When behind as reflectors they

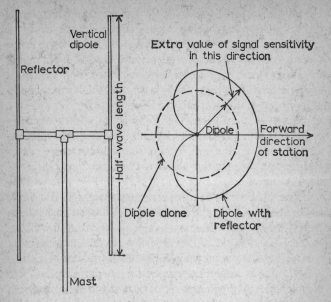

Fig. 146 Dipole aerial with reflector and polar diagram (in horizontal plane)

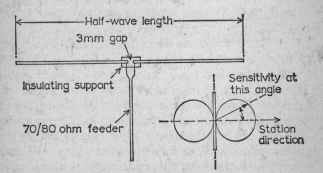

Fig. 147 Half-wave dipole horizontal aerial, showing how feeder cable is connected at centre, and polar diagram

are a little longer than the dipole and when in front a little shorter, the spacing being a quarter wavelength from the dipole. The effect is to make the dipole more sensitive in the direction of the required signal and less sensitive in other directions where interference may be located. Fig. 146 shows a dipole aerial (H-type) with reflector and the resulting polar diagram, which indicates the relative sensitivity at different angles around the dipole. It will be seen that, while the reflecttor increases the signal gain in the direction of the transmitting station, it cuts out entirely any unwanted signals or interference within a fairly wide angle in the opposite direction. A horizontal dipole is illustrated in Fig. 147, and the polar diagram shows that the response at right angles to the required signal is negligible over an appreciable angle. With the higher ranges of frequency, more parasitic elements can be used with advantage. There are combined aerials for receiving more than one band on the same aerial array and various shapes of aerial designed for special purposes, but they are seldom as efficient as the single-band aerial designed and used for its own purpose. Fig. 148 shows various outdoor aerials commonly in use at the present time, but it is not possible to show all the varieties and combinations obtainable, nor to deal with the theory of their design more fully in this work, for which reference should be made to several good aerial books that are now available.

Mention should be made of the frame or loop aerials with from three to eight turns of wire, spaced apart on 2·4 to 0·9 m square frames and having natural wavelengths of 160 to 185 m respectively. This aerial is directional with a figure-of-eight polar diagram, with a maximum response in line with the turns. Small frame aerials were once made for portable radio sets, but they have been largely superseded by the ferrite rod aerial for medium and long waves. The core material has a high magnetic permeability; it is responsive up to about 100 MHz, and very sensitive aerial coils on a ferrite core can be provided in a very small space for v.h.f.

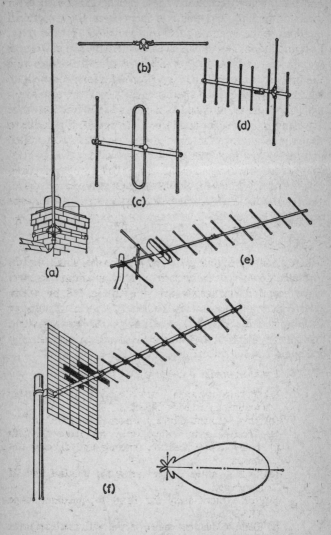

(b)

(d)

(c)

(a)

(e)

(f)

and u.h.f. signal reception. These aerials can be used for direction-finding, and this is evident from the directional qualities of the small portable transistor sets that use them. Loft and indoor aerials can be used where the signal strength is sufficient, and 7 to 10 m of insulated stranded copper wire hung in a roof space can be very effective for long and medium waves; but with limited space smaller aerials are also available, and these are designed to pick up a useful fraction of the wavelength. In areas of good signal strength, it is possible to use a simple wire as a f.m. dipole fixed on a suitably orientated picture rail in a room, with a piece of ordinary twin parallel (not twisted) flexible lead to the set for Band II f.m. reception. The overall horizontal length of the dipole should be 1·59 m, and the lead should be connected at the centre where the wire is cut into two halves.

Downleads

Where signal strength is good and interference small, a plain insulated downlead can be used for long, medium and short wave reception with aerials such as (a) in Fig. 148, but where interference is troublesome the use of special feeders or downleads and matching transformers may be necessary. For all the television bands it is necessary to use specially

---

Fig. 148 Typical outdoor aerial arrays

    (a) 5·5 m skyrod aerial for long, medium and short wave reception, lashed to chimney

    (b) Single-element dipole, horizontal, for Band II

    (c) Three-element vertical array for Band III with folded dipole (centre), reflector and director elements

    (d) One + seven element array for Bands I and III combined

    (e) Ten-element array for Band IV, medium range (10–20 miles)

    (f) Band V thirteen-element aerial with mesh reflector and polar diagram

designed downlead cable in order to increase the aerial gain and minimise interference, which is more troublesome in the higher frequency bands.

The aerial and its feeder cable must be taken together in considering the natural wavelength or tuning resonance of the aerial installation. The old long-wire outdoor aerial of the early years of radio broadcasting, with a length, including the downlead, of about 30 m, would have a natural wavelength of about 120 m (the Post Office maximum permitted length is about 46 m, or 150 ft). The receiver tuning device is then combined with the aerial and feeder to adjust the overall resonance to match exactly the wavelength of the station to be tuned in. The downlead can pick up interference that is missed by the aerial, such as radiation from the ignition systems of cars and main wiring in the house. The feeder cable is therefore an important part of the aerial installation, and for best results must be matched to the electrical characteristics of both the aerial and the receiving set. This is done by comparing the electrical impedances of each item and matching the values as exactly as possible.

The impedance at the centre of a half-wave dipole aerial is about 72 Ω, and in this country feeder cables are designed to match this value. The two common types of cable are the twin, balanced cable and the co-axial cable. Fig. 149 shows these two types of cable. The impedance is a function of the diameters of the inner conductor and the inside of the outer conductor in the case of the co-axial feeder, and a function of the conductor diameters and their spacing in the case of the balanced feeder. The co-axial feeder is mostly used because it is naturally screened by the earthed outer conductor, is less affected by local interference and is more robust. The balanced feeder is more liable to damage, its characteristics are liable to vary with a wet surface in bad weather and it is affected by local interference unless it has a metal outer screen like the co-axial, but this makes it much more expensive.

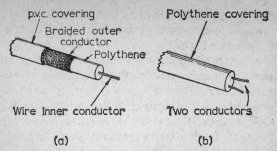

Fig. 149 Construction of a typical co-axial feeder is shown in (a),
a balanced line appearing as in (b). Screened balanced
cable is similar to (b) but has a braided outer screen and
covering as in (a)

The next important characteristic of feeder cables is
attenuation, which is loss of signal strength due to transmission losses in the cable and which increases with frequency. It is expressed in decibels (dB), which is a unit based
on the smallest sound audible to the human ear. It is used as a
ratio of signal strength, based on voltage or power, between
noise or interference and required signal, or, in the case of a
cable, between signal strengths at the receiving set and aerial
connections, and it is also applied to loss and gain of signal
strength in aerials, amplifiers and other audio/radio equipment. Tables are available that show numerical values of
voltage and power loss or gain against decibel values.

Noise, as distinct from other forms of extraneous interference, is inherent in all audio equipment—in particular, the
receiving set—and is apparant as a hissing background to the
required signal from the loudspeaker. The signal/noise ratio
must be kept as high as possible and, for ideal conditions,
should approach 50 dB.

Installation accessories
Needless to say, outdoor aerial fixings and aerials must be
rigid, secure and strong enough to withstand high winds,

snow and ice. The disposition of an aerial, as regards both inclination and direction, is important for maximum signal strength; therefore, if this is affected by weak fixings or weather conditions, the value of the aerial will be lost, and a loose aerial will cause variations in picture quality as well as signal strength.

A number of items of equipment for use with aerial feeders are available, including transistorised masthead amplifiers, power units and attenuators for controlling or regulating the aerial and feeder performance, interference suppression devices, matching or coupling transformers, diplexers and triplexers for connecting aerials of different bandwidths to a common downlead to ensure proper matching in each case, and star resistor networks and amplifiers to enable several sets to share the same signal from a common aerial. Nearly all these items aim to match impedances and to maintain signal strength under different conditions of use.

One device worthy of special mention is called a 'balun' (a corruption of 'balance' and 'unbalance'). This is used to correct lack of balance between a dipole, which is itself balanced, and a co-axial feeder, which is unbalanced and may adversely affect the characteristics of the aerial—particularly a high-gain aerial—and possibly be unsuitable for the receiver circuitry. It can be in the form of a dummy quarter wavelength of co-axial cable with shorted conductors, connected at the aerial end of the feeder cable, or be a specially designed unit with a balance-to-unbalance transformer for connection between the unsuited parts of the transmission line.

A common requirement in the house is for a television set to be able to be used in different rooms. If separate aerial feeders are to be connected in parallel, matching pads or splitters must be used to maintain the same impedance matching for each outlet. But this is not necessary if co-axial cable extension leads are used from the socket in room 1 to a socket in room 2, and possibly again from room 2 to room 3, each cable being plugged into the feed socket of the previous lead. This avoids mismatching.

Provision for the aerial downlead is not always made in housing, but the unsightliness of aerial leads trailing down roofs and stapled untidily to outside walls and window frames should make such provision a necessity. A 20 mm steel or plastic tube should be run as direct as possible from the aerial position, generally on the highest chimney stack or in the roof space, to the receiver position. If exposed above the roof tiles, the conduit should terminate in an inverted 'U' and bushed end to prevent the ingress of rain. At the receiver end, the conduit should terminate in a standard outlet box for a radio socket or sockets, as shown in Fig. 150; these are designed for the connection of co-axial cable. If desired, the conduit can pass through one or more similar outlet boxes, possibly located in first-floor bedrooms for alternative use on its way downwards. If these provisions are made when a house is built or before a later redecoration, the conduit can be buried or otherwise concealed without unsightly wiring.

Aerial reception is seldom required for ordinary domestic sound radio reception in good signal areas because most receivers are extremely sensitive, portable, battery-operated transistor sets which do not need external aerials, except in remote areas. But Hi-Fi (popular name for high fidelity) radio and television receivers for the v.h.f. and u.h.f. wave-

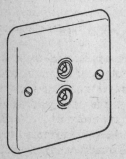

Fig. 150 Double T.V. co-axial socket outlet for direct connection to two separate T.V./f.m. co-axial downleads

bands usually require an aerial and a mains supply. The position is immaterial, but with television sets it is a good idea to locate them for viewing so that daylight from windows does not fall onto the screen, as this necessitates excessive picture brightness for adequate visibility. A mains outlet socket should be conveniently located for a supply to such equipment.

Earthing

The long-wire or long vertical rod aerial with a single conductor downlead requires a good earth connection for efficient oscillation to occur between the two poles (aerial and earth) of the resonant circuit through the coupling coil of the receiver. Indirect earthing through the set chassis or mains supply is very inefficient, although often quite effective in areas of good signal strength. The earth lead should be a stout copper wire of the shortest possible length.

Modern receivers are not always isolated from the supply by a double-wound mains transformer, and the chassis may be directly connected to the neutral line of the supply; further, the aerial system is only isolated from the set by series capacitors. Consequently, it is possible for dangerous conditions to exist with a 'live' aerial at mains voltage in the event of a fault or breakdown of certain components in the set. It is most important, therefore, to earth properly all exposed metal parts and the outer conductor of the aerial feeder cable, and to make sure that the supply neutral conductor is connected to the chassis if there is no mains transformer; it is also important to disconnect the set from the supply when adjusting the aerial or inspecting the chassis. In a.c./d.c. sets the chassis may be live and may have a separate earthing terminal.

Receivers

Little can be said of modern sound and television receivers that is not common knowledge; most makers produce sets of similar quality and performance, and best advantage of their performance can only be obtained by providing good aerial equipment, well sited and free of interference. The use of transistors and printed circuits has revolutionised receiver design. But, although transistors have increased the robustness of circuit construction and reduced the heat to be dissipated compared with that liberated by the thermal valves in older sets, there is still a risk of overheating and possible fire in the use of television sets if they are not carefully positioned where there is free ventilation around the cabinet and no likelihood of curtains accidentally covering the set.

Stereophonic broadcasts are also transmitted on Band II wavelengths, and these need decoding equipment at the receiver. The normal Band II aerial is used, but there is some loss of power and additional aerial gain may be required in weak signal areas. A 'mono' receiver will pick up the transmission, but if stereo reproduction is recquired a decoder will be necessary in the receiving set to separate the A-B sidebands of the carrier wave.

Colour television is transmitted on Bands IV and V on u.h.f. waves in most areas and, of course, requires a special colour receiver. Colour television sets are expensive—being three to four times the cost of monochrome sets—and repairs and replacements are also very much more costly, so colour television hire on a rental basis may be more attractive than purchase since the charges are not much greater than those for hiring a monochrome set at present. Bands IV and V aerials may be used, but they need to be more responsive at the edges of the bandwidths to cope with the extra subcarrier frequencies involved and to ensure good colour rendering. In general, colour television requires a better standard of aerial design and installation than is necessary for monochrome reception to ensure higher signal level and adequate bandwidth response, especially where signals are not strong. Colour television sets are also sensitive to masses of metal, such as radiators, in ther vicinity.

The theory and construction of radio receivers, amplifiers, tuners, loudspeakers and other electronic equipment for high fidelity sound reproduction are specialised subjects beyond the scope of this book, although such equipment is found in many homes; so, for more information in these directions the reader is referred to the many handbooks now available on these subjects. Suffice it to say that the electronic circuits in the receivers amplify and convert the radio signals to audible and visual effects in the loudspeaker and cathode ray tube respectively.

Telephones

The public telephone system is continually expanding as more and more households find the telephone a great convenience and within their means. The modern home should therefore have provision for a telephone service to avoid the unsightliness of telephone wires stapled to window and door frames, skirtings and walls. The simplest requirement for a new house is a concealed conduit from a point in the front of the house, outside and near ground level where the telephone cable is likely to enter, to a position inside the house where the telephone will be located. The Post Office engineers can then be directed to pass their wiring through the conduit, and the good appearance of the exterior and interior of the house will be preserved.

Although the telephone position is usually in the entrance hall, it can be in any room, and the conduit, which may be steel or plastic, should terminate in a standard flush outlet box containing a four-terminal connector block and having a cover plate with a bushed centre hole for the telephone cord, as shown in Fig. 151. It is not uncommon to have an extension telephone in another room—for example, a study or bedroom—in which case the empty conduit should extend from the outlet box at the first position to a similar outlet box in the second position. More than one extension can be provided if required, for which the conduit is extended in the same way. In such an arrangement the telephone instruments are usually connected in parallel on a single-line service and, if required, each instrument can have a push button to mute its own bell. There are also more elaborate systems with

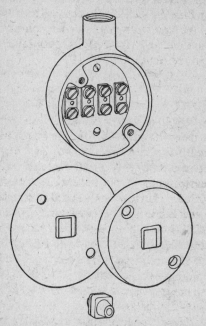

Fig. 151 Telephone outlet box with flush and surface lids and rubber bush for cord

central switchboards, but these are not usually required for domestic use.

Extension telephones or telephone socket outlets are a great convenience and often worth the small extra rental charge for each instrument or socket. The Post Office can provide telephone instruments in different colours to choice, and the latest designs have various toned sounders instead of bells. The telephone system voltage is low and wiring with low-grade insulation is used.

The old telephone exchanges with long lines of plug-in switchboards and operators are gradually being replaced by fully automatic switchboards having numerous rotating multi-contact selector switches and meters for registering each subscriber's calls. It will eventually be possible to dial a call on any telephone instrument for connection to any part of the world.

Although not usually required for home installations yet common in blocks of luxury flats, it is conceivable that the entrance telephone might become popular in the houses of a prosperous community of the future. This would be a call-push and loudspeaking telephone unit recessed in the wall by the front-entrance door and a bell with a telephone instrument in, say, the kitchen or living room, or possibly a bedroom for use when sick. Thus verbal communication with a caller is simple and safe without going to open the front door. If it is desired to allow a caller to enter, an electric door lock is fitted with a control push or switch near the inside telephone which, when operated, unlatches the entrance door and allows the door to open.

Telephone instruments

The instruments used in telephone systems comprise the transmitter, which converts sound waves into electric currents for transmission over long distances; the receiver, which may be an ear-piece or loudspeaker for converting the electric currents back into sound waves; bells, buzzers or light signals for drawing attention to the need to answer a call; amplifiers and repeaters for loudspeaking telephones and long distance communication; and manual or automatic switching arrangements to connect callers to the required receiver.

Public telephone systems with their elaborate and complicated exchanges, where the main switching is carried out, are beyond the scope of this book, which will therefore be limited to explaining the basic principles of telephony.

The transmitter. Sound transmitters or microphones incorporate a diaphragm which is designed to vibrate in response to sound waves carried by the air and impinging on the diaphragm, so transmitting the frequency and amplitude, or pitch and loudness, together with the quality of the speech or music being communicated, to a mass of carbon granules filling the space behind the diaphragm. The diaphragm forms the front of a rigid insulating box containing the granules, and the diaphragm and a backplate behind the granules form the two electrical connections, so that a current passing through the carbon granules is varied by the changes in resistance caused by the varying mechanical pressure on them as the diaphragm vibrates with the sound. A section of a typical microphone is shown in Fig. 152. There are various designs of microphone, but they are usually constructed as a capsule, complete with terminals, which is encased in a telephone handset.

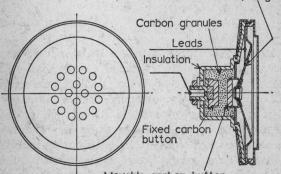

Fig. 152 Microphone capsule in section (leads shown diagrammatically)

A telephone induction coil is a transformer with an open magnetic core of soft iron wires, with the primary winding in series with a battery and transmitter, and the secondary winding to step up the voltage for the transmission line to the receiving end, where long lines are involved.

The receiver. The telephone receiver was originally an electro-magnet with an acoustic diaphragm of iron as the armature, which was separated by a small air gap from the soft iron pole-pieces of a permanent magnet. In the modern version of the receiver, the construction is much more efficient; the armature acts as a lever, pivoted at its centre across the poles of the electromagnet, and vibrates a conical diaphragm through a pin fixed at one side of the lever, as shown in Fig. 153. This unit is encapsulated in a similar way to the transmitter for enclosing in a moulded handset. The electric current, corresponding to that in the transmitter, passes through the coils and causes the diaphragm to vibrate in

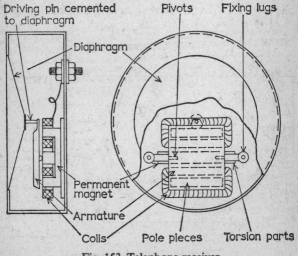

Fig. 153 Telephone receiver

unison, thus converting the electrical energy back into mechanical energy and the varying current into sound waves, which reach the ear.

If two such receivers are connected in series, a simple telephone circuit is formed. One will act as a transmitter and will generate an electric current due to the voltage induced in the solenoid coils by the varying magnetism effected by the vibrating diaphragm. The other receiver will act as a receiver in the normal way. While this arrangement can be easily demonstrated, the power and current are too small to be of use in practice without a battery in circuit to provide a larger power supply to the circuit.

The telephone circuit

A complete telephone circuit is shown in Fig. 154. Automatic switching is obtained by hanging the handset on a lever which operates the two-way switch, leaving the bell circuit in series with the distant station when out of use and putting the handset in series with the distant station when it is lifted off the hook; and when the call push is operated before lifting the

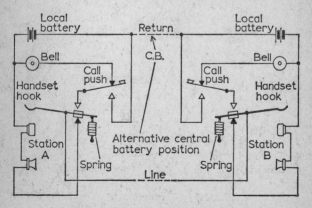

Fig. 154 Complete telephone circuit

handset it rings the distant bell. This is suitable for domestic installations, but for long distances induction coils would be connected in the circuits, as shown in Fig. 155. In a domestic circuit, very little power is required, and a few Lechanché or dry cells are adequate to overcome line and instrument resistances; in large installations, however, batteries of secondary cells up to 100 V are often used.

In the field work and for very long distances, economy in

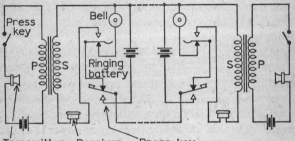

Fig. 155 Complete telephone circuit with induction coils

battery power is obtained by using magneto ringing for the bells. A magneto is an a.c. generator, geared for hand turning, and is automatically combined with the ringing key and connected in place of the battery to power the bell circuit. The bell is then a polarised bell with a capacitor in series, and the construction is similar to the a.c. mains bell described later (see page 266).

Wiring sizes are, of course, related to voltage drop and circuit resistances, but in small domestic installations 1·0 mm^2 or 1·5 mm^2 copper bell wiring can be used with battery voltages up to 15 V; for higher voltages compliance with I.E.E. Wiring Regulations is involved. In the best installations concealed 16 mm or 20 mm conduit would be used for wiring, but as twin bell wire is so small and not very noticeable if fixed with insulated staples on surfaces in a neat and inconspicuous manner, it is generally installed in this way if a higher standard is not called for. This wiring must be completely separated from any other mains wiring.

Loudspeaking telephones

With the development of transistor circuitry, there is now a trend towards the use of loudspeakers for telephone receivers and sensitive microphones without handsets, but there will always be a need for the handset telephone instrument where private conversation is necessary. The new trend, however, has many advantages: it eliminates the handset, which cannot be constructed on very hygienic lines; the microphone and remote loudspeaker are ideal for monitoring baby welfare in the nursery, and also allow freer movement and keep the hands free for writing while conducting a conversation.

Loudspeakers are simply a larger version of the electro-magnetic receiver, having a large conical diaphragm to move a greater volume of air; this requires more power to operate than the handset receiver and therefore amplification of the signal is necessary. Before the advent of the transistor, this meant expensive equipment with thermionic valves which made it generally impracticable, but the comparative cheap-ness and small space requirements of transistors and printed circuits have made amplification of very small electric currents feasible, especially for audio purposes.

Bells

The d.c. trembling bell is a simple piece of electromagnetic apparatus; the working principle is illustrated by the skeleton circuit diagram of Fig. 156. The electromagnet consists of a soft iron core, C, which is easily magnetised when the current traverses the magnet coils, MM; but when the current ceases the iron loses its magnetism. One end of the magnet coils is connected to terminal T, while the other terminal, T_1, is connected to an insulated brass pillar, P, which carries an adjustable contact screw, W. The tip of the screw makes contact with the spring, S, and the circuit is completed by pressing the bell-push. The current through the winding, MM, causes the iron core to be magnetised and the armature, A, is attracted. This breaks the contact with the spring, S, the current in the coil ceases, the magnets lose their magnetism and the armature flies back. The action is then repeated. This

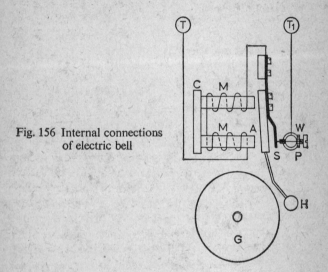

Fig. 156 Internal connections of electric bell

gives a trembling movement to the hammer, H, against the gong, G. Various types and sizes of gong, either of bell metal or steel wire, can be obtained to give distinctive notes; but if a quieter note is required, a buzzer can be fitted. Sources of trouble in a bell or buzzer are generally due to a fault in the contact-breaker, caused either by weakening of the spring or by corrosion of the contact screw. Sometimes a breakage may be found in the internal wiring of the bell.

A buzzer works on the same principle as the trembling bell, but the hammer and gong are absent. The vibration of the contact-breaker, which is made relatively light, gives the characteristic high-pitched note.

Bells for a.c. are designed to vibrate with the supply frequency (50 Hz) and do not need a contact-breaker.

A good-quality bell or buzzer should be obtained with the components mounted on a base with a well-fitting cover to keep out dust. Bells of the bakelite-moulded pattern are common and are made suitable for either extra-low or low voltage operation (up to 50 V or 250 V respectively). In the latter case, better insulation is required and the contact-breaker must be more robust.

Continuous-action bells

These are bells which continue to ring after the distant bell-push or contact is released and only cease when the operating cord is pulled or the local battery is exhausted. The local battery circuit is shown in Fig. 157, together with the additional mechanism. With the first movement of the armature, A, the trigger, U, disengages from the projection on the end of the armature and makes contact with the auxiliary pillar, X. The bell is then directly connected to the local battery, as will be seen from the circuit diagram. When the operating cord is pulled down, the trigger again engages with the projection on the end of the armature, the original connection is restored and the local circuit is broken.

Chimes

A popular modern sounder for the house is the chime, which is a very simple device consisting of an electromagnet in series with the door-push and battery or transformer. When the circuit is completed by operating the push momentarily, the horizontal sliding magnet core of soft iron with nylon ends is drawn rapidly into the magnet coil against the pressure of a weak coil spring; the leading end hits the sounding metal and rebounds with the aid of the spring to hit another piece of sounding metal at the other end of its travel. The two pieces of metal are usually of different tones to sound like a chime and may be hanging tubes of different lengths or small strips of metal as used in a toy xylophone.

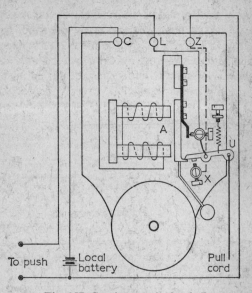

Fig. 157 Continuous-ringing bell

Bell-pushes

Bell-pushes of many types are on the market. For outside situations the weatherproof, metal-barrel type is best, while for bedside positions the ceiling pendant pear push or cord-operated pull type may be more convenient than the wall type and avoids the wiring having to be buried in the walls. For special purposes, such as burglar alarms, door, window and floor contacts are used, preferably with concealed wiring, but such items are usually provided and installed by specialist contractors. Fire alarm contacts which make the bell circuit when the front glass is broken are also a form of bell contact with reverse action, but fire alarms are not usually installed in the average house.

Relays

A relay is used to close a local battery circuit near the bell when the current from a distant push would not be strong enough to ring the bell, owing to the loss of voltage over a long run of small wire. Relays are also used with luminous call systems, the signal being sent by the distant push, which rings the bell, and the locking relay connects the local battery to the lamp, which remains alight until reset. The circuit diagram is given in Fig. 158.

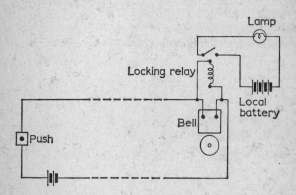

Fig. 158 Luminous call system with relay

The relay consists of a form of electromagnet with an armature which is similar to a bell mechanism but designed to have a single movement and carries contacts that make or break circuit with fixed contacts. They are also used with mechanical indicators in large establishments with long runs, as shown in Fig. 159, which shows a four-way indicator circuit with separate relays and a local battery.

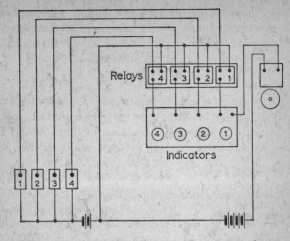

Fig. 159 Four-way indicator with separate relays

Electric bells on a.c. supplies
The magneto bell is operated on low-frequency alternating
current and its use is confined to telephone installations. The
construction is similar to that shown in Fig. 156, but the
contact-breaker is omitted; the two ends of the magnet coil
windings are taken directly to the terminals, T and T_1; the
armature, A, forms a rocker pivoted at its centre; and the
hammer, H, vibrates between two bells. With 240 V, 50 Hz
supplies a bell transformer is generally used for the bell
circuit. *This should be of the double-wound type*, in which the
two windings are entirely separate and insulated from one
another, and, if not double-insulated, have an earthshield
between them or on separate limbs of the core which must be
earthed. Fuses should be fitted to protect the primary (240 V)
winding, and one terminal of the secondary winding and the
iron core should be earthed if a metallic shield is not fitted
between the windings. With two bell-pushes, one at the front
door and the other at the back door, the low-voltage winding

can be connected via the respective pushes to a bell and a buzzer instead of using an indicator, as illustrated in Fig. 100, page 143, where the bell transformer, bell and buzzer are shown mounted above the kitchen door.

Indicators

Indicators are provided to show from which room a bell has been rung. There are three types of movement: pendulum, mechanical and electrical replacement. There is also the luminous indicator. Of these, the first is the simplest and consists of a pendulum with an iron armature which is attracted by an electromagnet and starts swinging. This is illustrated in Fig. 160, which also shows the circuits using a bell transformer. The mechanical-replacement movement allows a flag to drop and become visible and has a resetting rod projecting from the side of the indicator case, while the

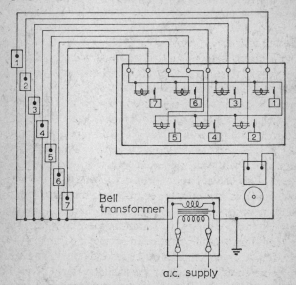

Fig. 160 Bell transformer and indicator circuits (supply neutral not earthed)

electrical-replacement type is similar but the flags are replaced by an electromagnet and a push, which can be located either adjacent to the indicator or at a convenient point some distance away. Some indicators include a bell, which is mounted on the base of the indicator. Luminous indicators employ single or groups of lamp signals connected in a similar way to that shown in Fig. 159, page 266. They are particularly useful where silence is required—for example, in hospitals—but can be used in conjunction with audible signals if required.

Bell wiring

Good materials and careful installation are essential if the bell system is to be reliable. Concealed conduit or surface wiring can be used. Similar general considerations as for light and power wiring apply. Wiring Regulations, however, do not apply to installations using a voltage less than 15 V, and lighter, smaller and cheaper materials can be used. No wiring should be buried solid in walls without protective tubing. The size of copper wire should be not less than 1/0·85 mm; single p.v.c., lightly insulated, in seven different colours, and twin bell wire is used for internal wiring, but if single wires are used do not forget that 'lead' and 'return' wires will be necessary. Kinks in the wire must be avoided, otherwise the single-strand conductor may break in the future. For the best-quality work, especially in damp situations and outdoors, twin lead-covered or p.v.c.-sheathed wire should be used. For such situations and larger section wiring where voltage drop has to be allowed for in long runs, mains voltage cables as used for lighting are often employed. Brass saddles or plastic clips which do not cut into the insulation are preferable to insulated staples; but if staples are used they should have an insulating saddle made of fibre, with a well-rounded top to the staple. Bell wiring must be kept separate from lighting and power circuits, and must not be run in the conduit carrying these circuits.

There are many other arrangements of bell circuits designed to suit special requirements, and when there is a considerable distance between bell and push an 'earth return' is sometimes used to economise on wire, using earth plates or water pipes for the return connections. Gas pipes must never be used for earthing connections.

For signalling purposes, single-stroke bells and buzzers are used; special contacts which operate like a Morse key can also be used, and such circuits often emply a single-line wire and earth return.

12 Electrical Safety and Treatment for Shock and Burns

Although the use of electricity in the house has developed rapidly since the beginning of the century, electrical accidents are still too frequent and serious to neglect some mention of safety considerations and how to deal with electric shock.

Safety

First, the provisions for safety made by the electrical industry should be recognised. The Institution of Electrical Engineers publish their *Regulations for the Electrical Equipment of Buildings*, which are designed primarily to ensure safety from fire and shock, and are accepted throughout the industry as the main guide to good practice. These ensure good insulation, construction, earthing and installation methods, which only need the addition of good workmanship to make an installation quite safe to use. Good workmanship is encouraged by having a National Inspection Council with a Roll of Approved Contractors, who undertake to carry out safe installation work and the employment of graded and qualified electricians.

Beyond this, the user must be discriminating in purchasing electrical appliances if the chain of safety is to be maintained. To help him there is a national organisation called the British Electrical Approvals Board, which is supported by the Electrical Industry and the British Standards Institution and is intended to test and approve domestic electrical appliances, both imported and made in this country, for safety. Therefore, any appliances or equipment bearing the B.E.A.B. stamp, shown in Fig. 161, can be relied on for safe use. The materials and equipment used in wiring installations are covered by various British Standards which ensure the safety, suitability and adequacy of these items, and British Standard Codes of Practice cover proper installation design (in particular, C.P. 324.201: Installation of Domestic Electric Space Heating Equipment; C.P. 324.202: Domestic Electric Water Heating Installations; and C.P. 321: Electrical Installations). Other Codes of Practice of relevant interest cover Supply Intake Arrangements; Private Electric Generating Plant; Lightning Protection; Bell and Call Systems; Earthing; Reception of Sound and Television Broadcasting; Radio Interference Suppression; and Telephone and Telegraph – Private Services.

Fig. 161 B.E.A.B. mark

In the past, many electrical appliances have been imported from abroad that were poorly constructed and dangerous to use, and some have caused fatal accidents. One of the most dangerous hazards has been the colouring of flexible lead cores, Continental standard colouring being different from British standard colouring, but this has been overcome by the adoption in this country of International standard colours, which is now compulsory under the Consumer Protection Act 1961. It is therefore most important for

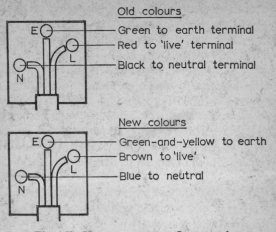

Fig. 162. How to connect a flex to a plug

safety to know the correct colouring, as shown in Fig. 162. In this respect the following safety hints should be followed:

1. Follow the directions or polarity markings in the plug.
2. If the flexible lead core has other than standard colours, do not connect it without advice from a qualified electrician, or an electrical shop or showroom.
3. Never use a two-pin plug for a three-wire flex.
4. Never use the earth terminal when connecting a two-wire flex to a three-pin plug.
5. If the appliance has a metal case, a three-core flex and three-pin plug must be used unless the appliance is double-insulated and marked ▣.

In using electrical appliances, it is important to remember that heat from the appliance must be allowed to escape, and that restriction of ventilation or cooling will cause the appliance to get overheated and may result in a fire. This applies to all electrical equipment, including T.V. sets and light fittings. Even electric blankets will cause fires if covered by too much for too long. All exposed live electrical elements, such as in electric fires, must be properly guarded. Flexible leads must never be used when damaged or when the insulation is old and hardened.

Long, trailing flexible cords are a source of danger and should be avoided if possible—extension leads can be used when necessary, but they must be fitted and connected properly so that only protected sockets (not projecting pins) of couplings are live when plugged into an outlet.

Flexible leads can easily become damaged if laid under lino or carpets, or if they are passed through doorways or windows. Always remove plugs when cleaning portable appliances, or switch off the control unit in the case of a cooker. Never take a portable appliance into a bathroom.

Make sure that any portable appliance is either all-insulated (with no exposed metalwork that can become live in the event of a fault) or, if not, that it is properly earthed. Earthing of the metalwork in an appliance ensures that it cannot become live at a dangerous voltage (see pages 92–5).

Electrical accidents

This subject can be divided into:

    (a) rescue of a victim of electric shock
    (b) treatment of shock
    (c) treatment of electric burns.

Much can be said under each heading, and the subject has been fully treated in books and papers; but in this work we can only skim the surface of modern knowledge in order to give the reader some insight and understanding of the situations involved. It is quite definite, however, that these situations require quick, cool and clear thinking. Ignorant and thoughtless action can be dangerous for all concerned.

Rescue

As the victim may be in contact with live conductors, the rescuer must not touch the victim immediately, and the first thing to do is to switch off or unplug the supply, if this can be done without delay. If not, the rescuer must insulate himself from 'earth' by standing on dry timber, using dry rope, timber or fabric to reach the victim, or remove him or her from contact with live conductors. The victim should be moved if necessary to a place of warmth and safety, using blankets if possible to keep him warm.

Shock treatment

Treatment of shock is most urgent because it has two vital effects: it may stop the heart's action directly or it may stop respiration, both of which must be restored as soon as possible. The most common situation is for a live appliance to be held in one hand while the other hand touches earthed metal or while standing on a damp concrete or tile floor. The effect is to contract the forearm muscles and tighten the grip so that the victim is unable to let go. Another effect of accidentally touching a live conductor is for the shock current to cause rapid withdrawal of the arm, and in passing through the body, the back muscles contract and throw a person over backwards. The passage of shock current through the body from one upper limb to some other limb generally affects the heart by causing ventricular fibrillation or disturbance of the normal action of the main pumping chambers of the heart, and this is usually fatal. Respiration is affected if the shock current passes into the head and through the respiratory control centre at the base of the skull to one of the limbs, or the shock current may cause contraction of the respiratory muscles and result in asphyxia.

If the heart is affected, blood circulation can be maintained by cardiac massage—by rythmical squeezing of the chest between breastbone and spine—but this is a dangerous practice in itself and is not considered safe enough as first-aid except by experienced and qualified first-aiders. In the case of arrested breathing, however, artificial respiration will often save life by introducing the necessary oxygen into the system and relaxing the respiratory muscles. Research has found that with the average healthy adult the shock voltage has to be over 40 V to pass a shock current of more than 40 mA through the body, which is dangerous, especially if this persists for more than 100 milliseconds; so electrical protective devices are usually designed and installed to operate on this basis. Painful shocks are nevertheless felt at lower voltages down to 25 V, and even lower voltages than this are dangerous to animals.

Artificial respiration. It is clear that urgency demands immediate action in an effort to restore normal bodily conditions. As soon as the victim is clear of contact with electricity, any false teeth, vomit, etc., should be removed from his mouth and artificial respiration must be started at once and continued unceasingly until normal breathing is obtained—even during a doctor's examination and the dressing of wounds if possible.

There are two approved methods of artificial respiration: the Silvester method and the Mouth-to-Mouth method. In the first method, the patient is laid flat on his back on a firm surface with the shoulders raised on a cushion or folded blanket and his head falling backwards. If necessary, the head can be turned sideways to help to keep the mouth clear. The rescuer kneels astride the patient's head, grasps his wrists and crosses them over the lower part of the chest. The rescuer then rocks his body forwards, pressing down on the patient's chest, and, releasing the pressure, draws the patient's arms backwards and outwards as far as possible with a sweeping movement. This causes air to flow in and out of the lungs, and also affects a degree of cardiac massage. This procedure is repeated about twelve times a minute with adults and up to twice this rate with children. Meanwhile, the mouth must be kept clear.

The modern approved method which has superseded older methods where there is no head injury is the 'mouth-to-mouth' method, popularly called the 'kiss of life'. The air

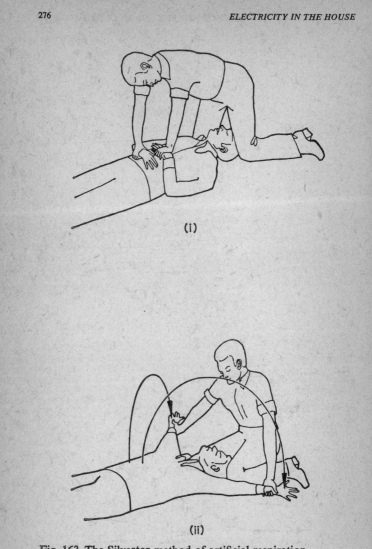

(i)

(ii)

Fig. 163 The Silvester method of artificial respiration

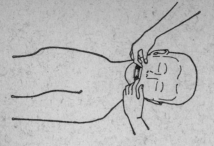

(i)

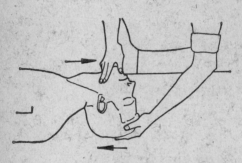

(ii)

(iii)

Fig. 164 The mouth-to-mouth method of artificial respiration

passages are cleared by tilting the patient's head back gently; the nostrils are closed by pinching with forefinger and thumb (in the case of adults only) or pressing the cheek against them while operating; and, with the fingers of the other hand holding the chin of the patient from underneath to keep the mouth open, a deep breath is taken and the mouth placed over the patient's mouth so that there is no leak of air. Then, after breathing into the patient's mouth gently—especially so with children—until the chest rises, the mouth is taken away to allow air to be expelled and the process is repeated every five seconds. The important thing about artificial respiration is to keep it up continuously until the patient has recovered normal respiration or until a doctor advises that treatment can cease. Colour returning to the face is an indication of recovery and is often evident—particularly with women and children—before self-breathing takes places. With the help of others in relay, a long period of treatment can be achieved. But one should always use the method one knows best, whether old or new, if success is to be ensured.

Resuscitation. Several methods of causing spontaneous breathing besides artificial respiration have been tried, such as the use of certain drugs, oxygen and carbon-dioxide, and special face-pieces for application, but they require special knowledge and equipment which can never be at hand when urgent first–aid is required.

Burns

Old-fashioned methods of treating burns with various house-hold substances as first-aid treatment are most unsatisfactory, and the modern first-aid treatment of burns is to keep the affected parts covered with a sterile dressing or a clean hand-kerchief soaked in cold water until proper treatment by qualified persons is obtained. Submersion of the affected part in cold water or other non-flammable liquid for a period will assist recovery. These methods alleviate pain and prevent contact with oxygen in the air, which is the essence of burn treatment.

Delayed shock

Physical shock and collapse can occur minutes after apparent recovery, so the patient should be watched and cared for with stimulants, if necessary, for a while before being finally released.

Index